Intention and Text

CONTINUUM LITERARY STUDIES SERIES

Also available in the series:

Active Reading by Ben Knights and Chris Thurgar-Dawson
Beckett's Books by Matthew Feldman
British Fiction in the Sixties by Sebastian Groes
Canonising Hypertext by Astrid Ensslin
Character and Satire in Postwar Fiction by Ian Gregson
Coleridge and German Philosophy by Paul Hamilton
Contemporary Fiction and Christianity by Andrew Tate
English Fiction in the 1930s by Chris Hopkins
Ecstasy and Understanding edited by Adrian Grafe
Fictions of Globalization by James Annesley
Joyce and Company by David Pierce
London Narratives by Lawrence Phillips
Masculinity in Fiction and Film by Brian Baker
Modernism and the Post-colonial by Peter Childs
Milton, Evil and Literary History by Claire Colebrook
Novels of the Contemporary Extreme edited by Alain-Phillipe Durand and Naomi Mandel
Recalling London by Alex Murray
Romanticism, Literature and Philosophy by Simon Swift
Seeking Meaning for Goethe's Faust by J. M. van der Laan
Sexuality and the Erotic in the Fiction of Joseph Conrad by Jeremy Hawthorn
Such Deliberate Disguises: The Art of Phillip Larkin by Richard Palmer
The Palimpsest by Sarah Dillon
The Measureless Past of Joyce, Deleuze and Derrida by Ruben Borg
Women's Fiction 1945–2000 by Deborah Philips

Forthcoming titles:

Beckett and Phenomenology edited by Matthew Feldman and Ulrika Maude
Beckett and Ethics edited by Russell Smith
Beckett and Death edited by Steve Barfield, Matthew Feldman and Philip Tew
Beckett and Decay by Kathryn White
Cross-Rhythms by Keren Omry

Intention and Text

Towards an Intentionality of Literary Form

Kaye Mitchell

continuum

Continuum International Publishing Group
The Tower Building
11 York Road
London SE1 7NX

80 Maiden Lane, Suite 704
New York
NY 10038

www.continuumbooks.com

British Library Cataloguing-in-Publication Data
A catalogue record for this book is available from the British Library.

ISBN: 978-1-4411-9864-8 (paperback)

Library of Congress Cataloguing-in-Publication Data
A catalog record of this book is available form the Library of Congress.

Typeset by Newgen Imaging Systems Pvt Ltd, Chennai, India
Printed and bound in Great Britain by Biddles, Kings Lynn, Norfolk

Contents

Preface: Intentions

At present, the weight of evidence against intention in literary theory very nearly amounts to a mathematical proof as to why, exactly, 'intention' is nothing more than 'fool's gold'. Just to be asked to even think about intention seems to be a request to trundle out arguments that have the air of redundancy. The very lack of current debate about it in literary academia would suggest that it has become an obsolete concept [. . .]. So whatever did happen to 'intention'?

Earnshaw 1996

What 'happened' to intention is so well known that it seems impossible to read Earnshaw's question, in the epigraph, as anything other than a rhetorical one: 'intention' went the way of 'the author' in literary studies, some time hence; its banishment has been compounded by a more general critique of the subject as the source of meaning, and the apparently irreversible linguistic turn within criticism – the triumph of critical approaches which respond to, without necessarily endorsing, a Saussurean way of looking at language and the individual. That this banishment has not been so successfully accomplished outside the academy – where authors and authorial meaning still hold sway – might, for a moment, give us pause for reflection, but it is easy enough to explain away this reverence for authorship as just one of the 'commonsense' or 'conventional' beliefs that it is precisely the job of theory to unpick.

Nevertheless, intention remains a *problem* within literary studies, despite and even because of the 'weight of evidence' against it; a problem the nature of which we have not quite mined. What I want to do in the pages that follow is to explore the problematic nature of intention, the history of its apparent exclusion, and to develop possible strategies for re-integrating this troublesome concept into our critical lexicon without thereby urging a return to some earlier, pre-theoretical vocabulary or some supposed golden age of authorial sovereignty. The use of this term 're-integration' should not suggest, however, that I am in agreement with Earnshaw's description of intention as effectively 'obsolete' or invisible or that I endorse his comment about the 'lack of current debate' on the subject; in fact the debate about literary intention continues, albeit mostly within the pages of the *British Journal of Aesthetics* in recent years[1] and mostly within certain rather restrictive and apparently impermeable parameters; it is these very parameters that I am keen to test. A case in point is Paisley Livingston's recent *Art and Intention: A Philosophical Study* (2005). Livingston, having already touched upon the subject of intention in an earlier book, *Literature and Rationality* (1991), here describes himself as a 'partial intentionalist' (rather than a 'strong' or 'absolute' one) and devotes a whole book to the elaboration of this position (2005: ix). Taking action theory (particularly that of Alfred Mele) as the basis for his argument, and asserting always the reasoned intuitiveness of his claim that intention is relevant whenever we take an interest in 'works as artistic achievements', Livingston yet rejects outright any kind of 'textualist intentionalism' (111). His book showcases, simultaneously, the very breadth of the topic (his discussion ranges across topics such as authorship, oeuvre, artistic value, interpretation, fictionality and collective intentions) and the narrowness of its treatment to date (in a manner rigidly in keeping with the dictates of analytic aesthetics).

Livingston's modest aim in his book is to keep the subject of intention 'on the research agenda in aesthetics' (211); it is my contention that intention, in some form or other, has never really gone away and that it merits a consideration outside and beyond the field of literary aesthetics. Intention has lurked, as suppressed premise or negative 'other', in a variety of theories of meaning (literary and non-literary) in the years since Wimsatt and Beardsley first appeared to send it into critical exile. We might, nevertheless, attempt to make it more at home in the present climate, less socially awkward and shambling and out of place in its old style authorialist garb. This welcoming of intention back into the fold of literary theory will require its wholesale redefinition, so we must from the outset be prepared to *re-think* intention.

Of necessity this 're-thinking' will be a polemical exercise, for the arguments over intention have been central to more fundamental and wide-ranging disputes about the place of theory within literary studies. As one of the staunchest opponents of Theory (capitalized to suggest its monolithic and sinister character), Valentine Cunningham has written:

> Under Theory, the text is demonized by a clamantly Pyrrhonistic rhetoric of lapse, failure, lack, disablement, deficiency. This rhetoric of deploring is all over the pages of Theorists. The text is in ruins, a ruin, a bombsite [. . .]. The text according to Theory is fragmented, bitty, broken. It can't speak out; it stutters; it hesitates; it can't see; it's blind; it's occluded. It's disfigured, defaced, an affair of de-facing, of de-personifying [. . .]. The text does not know itself; it's unconscious of what it's up to; it's repressed. It can't walk straight; it staggers; it's lame; it's maimed; it errs. The text misleads: it's labyrinthine; amazing; aporetic. (2002: 60)

One of the things that the text is seen to lack, of course, is the animating and semantically determining authorial intention, and thus its connection to the human. The job of this book, as I see it, is neither to restore to the literary text a sense of its own cohesion, stability or unity (or to re-establish this troublesome connection to the human), nor to contribute to its ruination (as Cunningham interprets it). Instead, my desire here is to reflect systematically upon (i.e., to theorize) intention, and to pursue a preliminary perception I have that, in the manner of smoke and fire, where we find meaning, we will also find at work some notion of intention, however schematic and undeveloped. We might then begin to reflect upon what is meant by 'intention', and to whom – or what – we can and do attribute intentionality in the literary context.

This project finds its motivation in two key intuitions: first, that the concept of intention, despite the acres of print dedicated to arguments which declare themselves 'for' or 'against' it (or perhaps even indifferent to it, although combatively so, in a manner that suggests this indifference cannot be absolute), remains a concept resolutely *untheorized*; second, that the task of theory (if it must have one) is precisely the kind of tracking, interrogation and elucidation of a concept so familiar and fundamental as to appear known. There is, to borrow Shklovsky's terminology for a moment, a certain amount of 'habitualization' and 'automatism' in our relationship to intention – we know it, but we cannot see it (1988: 20, 21); the theorization of this concept therefore becomes a kind of defamiliarization, and this, perforce, is the real and revitalizing task of theory.

A warning, then: this 'elucidation' of intention may, in effect, be anything but; indeed it may render the concept in question more obscure, less tangible although not, I hope, less useful for that. The structure of this book does not, and could not, represent a kind of progressive enlightenment; it is not the case that all will become clear at the end, but rather that a general murkiness will be seen to inhabit those aforementioned discourses around intention. So, before we begin to argue for or against the presence of intention, before we sort critics into their respective camps on the basis of their attitude towards intentionalism, we must first decide what exactly is being included in the remit and reference of this term 'intention'. What is it that Wimsatt and Beardsley deem inaccessible and irrelevant (1962) on the one hand and Hirsch calls 'the soul of speech' (1976: 90) on the other? Unless and until we agree upon this, we cannot begin to argue either intention's centrality or its marginality to accounts of literary meaning, and we cannot agree if it is indeed being excluded (or promoted or passed over or privileged).

Cunningham, in the eloquent but laborious anti-theory diatribe cited earlier, is also expressing displeasure at what he sees as the usurpation of the text (and literary criticism) by this new and false god, Theory; he ends his book with the claim that 'though reading always comes after theory, theory is inevitably the lesser partner in this hermeneutic game' (2002: 169). On the face of it, the absence of any literary references within this book might seem to represent precisely the turning away from the literary object that Cunningham warns against – the close readings to be found here are close readings of theoretical and philosophical texts. However, it is an interest in literature and, specifically, in how literary texts produce meaning(s) that has initiated this enquiry, although the question of literariness itself is not addressed here and the 'text' of the title can be read as having a quite general application; it seems to me that the distinction between the literary and the non-literary was problematized, if not actually elided, long before post-structuralism came on the scene. Theory, in the sense of theoretical writing, is therefore neither prior nor posterior to literature and one kind of text is not 'greater' or 'lesser' than the other.

Cunningham's primary objection, according to my interpretation of his self-declared 'manifesto', lies with the *application* of theory to literature – or rather the *imposition* of the former upon the latter – and the effect of this: a certain reductiveness in reading. This, I certainly set out to avoid; the kind of theorizing that follows is a reflection upon the process of meaning and understanding in literature which precedes and surpasses any particular readings of particular texts. Therefore I have seen fit to leave out any exemplary critical readings, the tokenistic presence of which would not seem to me to advance the discussion in any notable way. Crucially, what I am not attempting here is the construction of a full-blown critical methodology, in the style of a *Validity in Interpretation*; the 'intentionality of form' that I will propose, at length, is not a set of instructions for reading literary texts – but then neither do I happen to think that deconstruction posits such a set of instructions, although many would disagree with me here. Nor am I especially concerned with the resolution of interpretative disagreements past and future, as an across-the-board consensus is, in my opinion, far more damaging and detrimental to literary studies than even the most intractable divergences of opinion among critics may be.

This matter of the provision of a critical methodology, however, does raise the question of the true nature and responsibilities of theory: when Knapp and Michaels declare themselves to be 'against theory' they do this by way of arguing that intention is not a 'theoretical issue', that is, not something that can or should be theorized about, because it is simply – pre-theoretically, if you will – identical with meaning (1985a: 17). For them, the suggestion that we can decide to describe ourselves as intentionalist or anti-intentionalist critics is what 'makes theory possible because it creates the illusion of a choice between alternative methods of interpreting' and 'to be a theorist is only to think' – erroneously, in their view – 'that there is such a choice' (18). Intention cannot be 'chosen' or argued for, it is simply there, so intentionalism and anti-intentionalism are equally misguided, and, they conclude, 'as soon as we recognize that there are no theoretical choices to be made, then the point of theory vanishes. Theory loses' (ibid.). I am not, therefore, proposing that we 'choose' intention as some kind of interpretative criterion, only that we choose, at the very least, to talk about it in new and productive ways, and to test its malleability and utility as a concept. This does not mean, however, that the account of intention I am offering here follows the logic of Knapp and Michaels' account: even if intention is 'simply' there, it is not there in quite the way that Knapp and Michaels propose.

As the title implies, the issue of form is also central to my argument here. In the long and, occasionally, torturous discourse on literary intention, intentionalism has always been figured as incompatible with formalism; it is the formalist manifesto of New Criticism which first deems references to intention to be 'fallacious'. This book sets out to dispute that assumption of incompatibility, by questioning what we mean – and what we might yet mean – both by 'intention' and by 'form' in the context of literary studies, and by challenging what I'm going to call the *logic of outside and inside* out of which both concepts, in their traditional formulation, emerge.

In considerations of literary meaning since the mid-twentieth century, the concept of intention has traditionally been located outside the text and it is most commonly identified, by all parties, as some kind of mental state or authorial blueprint. Hence intentionalists argue on ethical or humanist grounds for the 'right' of the author to determine in advance and for always the core meaning of the work that he produces and often end in describing works as psychological entities, the particular expression of some particular subject. Anti-intentionalists, with rather more success, argue on the grounds of intention's inaccessibility or irrelevance, invoking the 'other minds' problem or some notion of the purity and autonomy of form in their defence, and warning against a description of texts as uncomplicatedly expressive (of the mind of the author). Both sides, though, are ultimately reliant on an idea of intention's exteriority to the text (both also, therefore, imply that there *is* such a bounded and determinable entity as 'the text itself', whether this text equals an Idea in the mind of the author-god, or a series of marks upon a page). Intention, then, is seen as something that is added to (that is to say, separable from) the form of the text: it is either prior (in the case of authorial intention) or posterior (in the case of some readerly intention/construal of meaning) to this text.

Against this logic of inside and outside – seeing such a logic, in fact, as already comprehensively dismantled by deconstructionists such as Derrida and de Man – I aim here to work towards an understanding of literary *form* which yet retains the concept of intention or intentionality as an explanatory notion, and so to situate intention 'within' the text, while also envisaging this 'text' as substantially expanded and displaying a certain fluidity in its boundaries (so this term 'within' is, admittedly, somewhat disingenuous). This working *towards* a new kind of formalism will also, of necessity, be a movement *away* from a conception of intention as a mental state, a blueprint in the mind of the author which exists prior to the text that it purports to structure and whose meanings it claims to determine. Instead, an attempt will be made, at length, to conceive of intention as a structural or linguistic feature of texts, a feature of the way that they themselves 'intend' meaning.

My route to such a conclusion – which is rather a kind of opening, the creation of an aperture for intention within our understanding of literary meaning – will take me on a journey notable for its wilful interdisciplinarity: first, via an introductory consideration of the original exclusion and exteriorization of intention by Wimsatt and Beardsley, which suggests that some stubborn trace of intention remains in the text, despite their best efforts to be rid of it; next, by way of a discussion of Hirsch's attempts to recuperate intention by invoking an ethical commitment to authors and so sealing the conception of intentionalism as necessarily *authorial* intentionalism and intention itself as prior blueprint; then, in a chapter which branches out into the realm of analytic philosophy, via a discussion of speech act theory and the literary text as 'utterance'; from there to a discussion of phenomenological theories of intentionality and readerly intention; and finally, to the post-structuralist musings of Derrida and de Man, which both sound the death knell of the blueprint theory of intention while also, in their borrowings from phenomenology, making room for just such a reconsideration of the relationship of intention and form that I am calling for.

Finally, a note about gender. It appears that for the vast majority of critics, theorists and philosophers writing on literature and/or intention, authors are unquestionably male, as are critics and readers. While not wanting to contribute to this consistent elision of the female from such discussions, I have reluctantly decided that authors and readers will indeed figure as 'he' throughout this book: partly in order to avoid anachronism when the theory under discussion uses 'he' in this way, and partly to avoid the kind of stylistic clumsiness which results from the use of formulations such as s/he or he/she. This book also, regrettably, leaves out the question of gender politics in another way: it is quite pertinent to argue, given Hélène Cixous's comments on the subject (in *Coming to Writing*, e.g.), that the blueprint idea of intention emerges out of a conception of meaning and rationality (otherwise known as logocentrism) which is undeniably masculinist (and so actually phallogocentrist). That you will not find such a topic discussed here is the result of a very necessary kind of selectivity, for the subject of intention, I have found, opens up a potentially interminable series of questions and presents the theorist with all sorts of tantalizing possible trajectories. My own trajectory, then, is simply one of many, but I hope that its *intentions* at least are now clear.

Note

1 See, e.g., Davies, 'Artistic Intentions', 148–62; Leddy, 'Iseminger's Literary Intentionalism', 219–29; Trivedi, 'An Epistemic Dilemma', 192–206; McGonigal, 'Metaphor, Indeterminacy, & Intention', 179–90; Price, 'Artificial Functions', 1–17.

Acknowledgements

My initial work on intention was done as part of a doctoral dissertation at Birkbeck, University of London, generously funded by the Arts and Humanities Research Council, so thanks must go first and foremost to the A.H.R.C. and to Professor Steven Connor who supervised that dissertation with characteristic flair, imagination and intellectual acuity. There is no doubt that the project benefited enormously from his magnanimity and expertise as a reader and from the constructive and engaged comments of my two examiners, Séan Burke and Andrew Gibson.

Warmest thanks are also due to Bob Eaglestone, whose encouragement led me finally to return to the topic of literary intention and seek publication of the book, and to Anna Sandeman and Colleen Coalter at Continuum for their assistance and professionalism in bringing that publication to fruition.

Too many people to mention by name have aided and abetted my work over the last few years: fellow research students at Birkbeck; colleagues and students at the University of Westminster and, more recently, the University of Manchester; friends, family and loved ones. Particular thanks, though, must go to the following, without whom my life and work would be much impoverished: Joe Brooker, Daniela Caselli, David Cunningham, Kate Graham, Jennifer Greitschus, Jason Hall, Kyoo Lee, Gill Partington, Laura Salisbury, Alex Warwick, Leigh Wilson, Andrew Wyllie and – finally – my ever-loving and supportive parents, Mitch and Fiona Mitchell.

Introduction

Beginnings: The Birth of a Fallacy

Intention-to versus intention-by

In the summer of 1946, W. K. Wimsatt and M. C. Beardsley published an article in the *Sewanee Review* entitled 'The Intentional Fallacy'. This has influenced all subsequent work on intention's role in the construction and interpretation of the literary text and on literary meaning generally, although in fact the fallaciousness of which it was the scourge had little to do with the concept of intention figured metaphysically or ontologically, and a great deal to do with the practice of criticism and certain alleged epistemic realities. Ironically then, 'The Intentional Fallacy' has achieved what it was never 'intended' to achieve: the uncritical eradication of intention from the vocabulary of literary critics and, latterly, literary theorists. In the light of this eradication it becomes a polemical question to ask what Wimsatt and Beardsley meant by 'intention' and what they intended by its exclusion, before considering how this troublesome concept of intention might be recovered and resituated in a manner not antagonistic to the present theoretical context.

Wimsatt and Beardsley argued, famously, that, 'the design or intention of the author is neither available nor desirable as a standard for judging the success of a work of literary art' (1962: 92). The key points of discussion arising from this much-quoted slogan of New Criticism are as follows: the *availability* of an author's intention to the critic or theorist (an intentionalist will say it is available, an anti-intentionalist that it is not), the *desirability* of making this appeal to intention (ditto) and the question of whether evaluative judgements can take this intention as some kind of critical criterion. However, it is instructive to consider *what it is* that might be available or desirable or a standard, and the nub of Wimsatt and Beardsley's argument (which is not an argument as such but rather a summation of the New Criticism's critical ideology) is to be found in their particular conception of intention, what it means to them and where it is located. I do not mean to suggest by this that their conception is idiosyncratic, in fact it will be revealed to be all too popular and pervasive.

By 'intention', Wimsatt and Beardsley mean 'what he intended', that is, 'the design or plan in the author's mind' (92). This intention is an entity existing prior to the work itself, a mental design or blueprint detailing how the artist or author wants the work to be or what he wants it to express. The mentalistic characterization, the connection with consciousness, allows anti-intentionalists at once to employ psychologistic arguments which dismiss intention as inaccessible or as, at the very least, subject to change in the flux of the creative process. Certainly, such a blueprint is difficult to locate in space or time. Is it completed, perfected, before the author even begins work on the text? Thereafter, how mutable might it be? If the vision of the work incorporated in the blueprint is the finished work, at what point can the work be said to be finished, bearing in mind authorial revisions, second editions and so on? Is it only finished if it matches or corroborates the original blueprint version? In other words, what degree of variation is permitted? Wimsatt and Beardsley's location of intention as logically and temporally prior to the work of art strengthens their case against it, allowing Wimsatt, in a later solo consideration of the question of intention, to characterize the 'moment' of intention as doomed to 'recede and ever recede into the forgotten, as all moments do', thereby proving that as poems do not themselves 'dwindle in meaning' (but rather accrue meanings), this moment of intention cannot be as significant as the intentionalist supposes (1976: 38). The intention is 'outside' the text; the text achieves a kind of independence from the originating intention, a life beyond it. However, it is only intention's conception as 'momentary' which permits the possibility of this 'beyond'.

According to the blueprint conception of intention, the relationship between 'design or intention' and work is analogous to that between cause and effect, and indeed Wimsatt and Beardsley allow that the 'designing intellect' is a '*cause* of a poem' while stipulating that, as such, it should not be treated 'as a *standard* by which the critic is to judge the worth of the poet's performance' (1962: 92). The cause in this instance is, notably, private and the effect that it 'predicts' is provisional, due to the possible limitations or opportunities which the chosen form of expression will impose or provide. Significantly, it is separable from the work which it intends, as a cause must be separable from its effect (while also displaying some link stronger than contingency, weaker than entailment). I will call this blueprint conception of intention *intention-to* and it should be distinguished above all from *intention-by*. This latter will receive a fuller definition in due course but, provisionally, it tallies more obviously with authorial *meaning*: it is what the author means by the semantic and syntactic arrangements that make up the final text of the work. The thesis that *intention-to* has no relevance to literary evaluations is uncontroversial but trivial: uncontroversial because it is difficult to oppose the view that literary criticism should concern itself with what the author actually did do (what is, in fact, present in the text) rather than with what he might have originally intended to do, but failed to execute (or changed his mind about); trivial because it does not take us any distance towards a theory of textual meaning or determinacy (or other endlessly pressing issues), but merely discounts an avenue of enquiry which would have discounted itself before long.

Michael Hancher offers a more developed analysis of intentions as they figure in the realm of the literary, dividing them into three categories: 'programmatic', 'active' and 'final' (1972: 829, 830, 834). Programmatic intentions correspond to Wimsatt and Beardsley's 'design or plan', they are 'approximate and generic' and may miscarry (829). In fact, a certain fluidity and lack of specificity are necessary in-built features of the programmatic intention, because, 'if one were ever fully specific, down to the last verbal detail, the text would already be completed – albeit in the author's head, rather than on paper' (838). The success of a programmatic intention lies in a finished work falling within the remit of the original blueprint, but this is irrelevant to evaluative considerations of the work (although not, perhaps, to evaluative considerations of the author, his foresight and fulfilment of purpose, which are separate and debatably non-literary matters). The irrelevance of any knowledge of programmatic intention, at least in facilitating an evaluative judgement of a work, is Wimsatt and Beardsley's primary contention, and I do not dispute it.

By 'final intentions' Hancher means 'whatever the author wishes to accomplish *by means of* his completed work', that is, an effect 'beyond the mere understanding of his meaning' (1972: 834, 835). Such motivations are beyond the text itself, and may include an author's desire for wealth or commercial success or for a particular ideological effect in the minds or behaviour of his readership. Similarly Close notes the distinction between 'intention as an objective lying beyond the boundaries of the action' and intention which falls within the boundary of action, that is, the intention *with which* someone does or says something (1972: 175). The latter is more relevant to literary criticism, which is quite rightly unconcerned with the author's ulterior motives, whether propagandist or mercenary; it is the latter which figures intention as analogous to 'meaning'. Kemp also makes reference to 'ulterior' intentions (as contrasted with 'immediate') and dismisses them as artistically irrelevant; however, he fails to come up with a definition of an intention which falls between the blueprint (that which is chronologically prior to the artwork) and the final or ulterior intention (concerning effects which are posterior to the artwork) (1964: 147).

This is the intermediary position occupied by Hancher's 'active' intentions which are 'intentions to be (understood as) acting' and 'characterise the actions that the author, at the time he finishes his text, understands himself to be performing in that text' (1972: 830). Such intentions, says Hancher, 'bear on the text and shape its "meaning" at the moment of completion' (ibid.). While this is subject to the pseudo-ontological criticisms already detailed (e.g., when does the moment of completion occur? At what point does a text go from work-in-progress to completed work?) it is, nevertheless, a workable definition of *intention-by*.

It should be noted at this point that the categories of programmatic and active intention may occasionally overlap or at least bear upon each other, and this is the result of genre considerations. For example, the programmatic intention to write an elegy (and to be understood as doing so, i.e., as fulfilling the expectations and conventional requirements of that genre) will involve an active intention – namely, the intention to be understood as sorrowful or elegiac – simply because forms such as the elegy, the ballad and so on, conventionally entail certain broad meanings. Despite this possible boundary confusion, Hancher concludes that 'meaning is a matter of active intention, and not of programmatic intention, by definition' (838). My contention, then, is that there is no evidence in 'The Intentional Fallacy' of hostility to such intentions and that the hostility to both programmatic and final intentions which *is* indisputably in evidence need not entail a similar critical exclusion of active intentions. In fact, Wimsatt and Beardsley's insistence here that the critic concentrate on what the poet succeeded in doing and look for evidence of intention which is 'effective in the poem' (1962: 92) reveals a tacit endorsement of active intentions, yet perversely they take this privileging of what the poet succeeded in doing to be a refutation of intentionalism – witness the claim that 'An eminent intentionalist in a moment when his theory repudiates itself [says that] "the poet's aim must be judged at the moment of the creative act, that is to say, by the art of the poem itself"' (ibid.).

Yet, in the light of Hancher's distinctions, this is *not* a 'repudiation' of intentionalism if intentionalism is geared towards the retrieval of *active* intentions. Wimsatt, when he revisits the question of the intentional fallacy some years later, seems amenable to a discussion of authorial intention if this is the author's '*effective* intention or *operative* mind as it appears in the work itself and can be read from the work', that is, if this is his active rather than programmatic intention (1968: 36). Hancher's distinctions indicate, at the very least, that the term 'intention' is disparately applied,[1] and therefore that it is necessary, in declaring oneself an intentionalist or anti-intentionalist, to specify what *sort* of intentionalist or anti-intentionalist one is. This is something that Wimsatt and Beardsley manifestly fail to do. Had they done so, it would be immediately evident that 'The Intentional Fallacy' was an attack not on intention *per se* but on the critic's reference to material external to the text and so, properly, on the *direction of inference* in interpretation: Wimsatt and Beardsley insist that what is 'outside' the text is critically relevant only insofar as it can be inferred from what is 'inside' the text; the 'fallacy' of traditional authorial intentionalism, therefore, is that it infers what is 'inside' from what is 'outside', that is, from details of the author's biography and the context of writing, from a speculative assumption of what his programmatic intention must have been. The allegation of fallaciousness, then, rests upon a particular understanding of the boundaries of a literary work, of what can properly be said to be its 'inside' and its 'outside' and of what constitutes the form of that New Critical staple, the text itself. It is the question of the relation of intention to literary form which is at the core of this book and which largely serves to structure its argumentative framework: my suggestion here is that this relation has, to date, been inadequately theorized.

Internal evidence versus external evidence

In 'The Intentional Fallacy', Wimsatt and Beardsley insist upon a distinction between the *internal* evidence for the meaning of a poem, which is public, and discoverable, 'through the semantics and syntax of a poem, through our habitual knowledge of the language, through grammars, dictionaries, and all the literature which is the source of dictionaries, in general through all that makes a language and culture'; and *external* evidence, which is 'private or idiosyncratic; not a part of the work as a linguistic fact: it consists of revelations (in journals, for example, or letters or reported conversations) about how or why the poet wrote the poem.' (1962: 97–98) However, although they claim that the public nature of internal evidence and the private nature of external evidence is a paradox 'only verbal and superficial', it is questionable *to what* exactly 'our habitual knowledge of the language' is *internal* (97). Certainly it is not internal to the poem, suggesting as it does that in order to understand a poem, a critic must already be in possession of some knowledge concerning the workings of language, and perhaps some knowledge of the broader context in which the work was created, its linguistic peculiarities or culturally specific references. It is not clear how this differs from the kind of biographical information which Wimsatt and Beardsley are attempting to dismiss from criticism and they blur the self-imposed boundaries still further in their delineation of a third or 'intermediate' category of evidence, concerning, 'the character of the author or [. . .] private or semiprivate meanings attached to words or topics by the author or by a coterie of which he is a member' (98).

Few of the varieties of intentionalism available insist on more than this privileging of the meaning 'attached to words or topics by the author', and it is in their acceptance of an intermediate form of evidence (in 'moderation') that Wimsatt and Beardsley come closest to endorsing a weak form of intentionalism. Witness their claim that:

> The biography of an author, his use of a word, and the associations which the word had for him, are part of the word's history and meaning [. . .]. The use of biographical evidence need not involve intentionalism, because while it may be evidence of what the author intended, it may also be evidence of the meaning of his words and the dramatic character of his utterance. (ibid.)

However, it is only possible for them to maintain that 'biographical evidence need not involve intentionalism' on the grounds of their particular conception of intention as prior blueprint, a conception which is, for the most part, irrelevant to textual meaning and its decipherment. As we shall see in the next chapter, the most popular forms of intentionalism choose to substitute 'the meaning of his words' for 'what the author intended', sometimes even claiming that the two are logically equivalent, so the separation of them here is questionable – particularly if it is noted that a philosophical definition of 'meaning' cannot proceed with reference to 'intention' (and vice versa) without incurring charges of circularity, because the two concepts are so closely linked, one thus presupposing the other; I will examine the inseparability of these concepts in greater depth in the chapters that follow.

Wimsatt and Beardsley's extension of the class of permissible evidence in interpretation to include this 'intermediate' category also indicates a certain equivocation as far as the delimitations of literary form are concerned; the very existence of the 'intermediate' suggests that the boundary between 'inside' and 'outside' is, at the very least, permeable, if not actually radically unstable. Nevertheless, they persist in their internal/external distinction, even while extending their account of internal evidence to include notes by the author: 'Whereas notes tend to seem to justify themselves as external indexes to the author's *intention*, yet they ought to be judged like any other parts of a composition' (103). Such a qualification serves to reveal how tenuous the distinction really is. In fact, the point of the distinction is to exclude references to the author's psychological state in order to privilege 'the way of poetic analysis and exegesis', which is 'the true and objective way of criticism', and to denigrate 'the way of biographical or genetic enquiry' (104). Wimsatt and Beardsley play upon an internal/external distinction, not because they believe that the boundaries of a work of art can be clearly delineated (if they can be, these critics make no attempt to do so), but as an indication of the economy of their critical method and its resistance to romanticism. Despite their advocacy of 'objectivity' and stated desire for 'more precise terms of evaluation' than those characteristic of the intentionalist critic, the internal/external criterion of relevance is deeply subjective (97). It represents a kind of critical asceticism – dictating that critical procedures should be unsullied by appeals to anything other than 'the text itself' – combined with the epistemological tenet that intention cannot, in any case, be appealed to, because it is something private, inaccessible, irrecoverable, ultimately unknowable.

However, externality need not entail irrelevance, and does not in all cases, as Wimsatt and Beardsley themselves concede in their admission of 'intermediate' evidence. Furthermore, intention of the non-blueprint variety need not be external to 'the text itself', and this is something which Wimsatt admits in 'Genesis' in his discussion of '*effective* intention'. Of intentionalism which centres on intention 'as it is found in, or inferred from, the work itself', Wimsatt writes that 'Obviously the argument about intention [. . .] is not directed against such instances – unless in an incidental and general plea for clarity in the use of critical terms' (1968: 26). What Wimsatt does not do is fully elaborate this idea of intention found 'in' the text itself or consider what the ramifications of this might be for our conception of literary form: what is it for intention to be 'internal' to a text? In what ways does the form of a literary work display or evince intention? Does this change our thinking about what literary intentions are and, if so, how?

One notable point is that while the blueprint conception of intention raises the spectre of artistic failure – the author's failure to fulfil his original vision – intention 'in' the work, or achieved intention incurs no such problems. Skinner writes, succinctly, that 'a writer [. . .] will normally achieve what he intends to achieve and will normally intend to achieve what he achieves' (1976: 214). The apparent triviality of this doesn't affect its credibility, although it does diminish its usefulness as a possible interpretative criterion. Nevertheless, it is a formulation which might appeal to Wimsatt, who postulates in 'Genesis' that one may infer the intentions of the author from the meaning(s) of the poem but may not infer the meaning(s) of the poem from the intentions of the author: 'If a poet sees red, he may well either write or not write a red poem. If he writes a red poem, it would seem to be a sound enough inference [. . .] that he has in some sense seen red' (1968: 16). In other words, as previously suggested, it is the *direction of inference* (from work to poet, rather than from poet to work) that is of key importance here; this is a much narrower thesis than has been assumed by most of the inheritors of Wimsatt and Beardsley's anti-intentionalist argument and it does not amount to a denial of the presence of intention in the text. Of course, the claim that intentions are internal to the text can also be used as an argument against intentionalism, as it implies that they do not, therefore, require separate consideration. A weaker intentionalism, though, need not insist upon separate consideration, but only upon the necessarily intentional nature of texts. What it might mean for a text to be 'intentional' is the question haunting the chapters that follow; only at the end will any kind of definition of the intentionality of form be attempted.

Returning to Wimsatt and Beardsley, however, it is evident that there is some difference of opinion between them, and this in part explains why it is that the article of which they are co-authors should lend itself to such contradictory reasoning on the matter of what is and what is not external to the text. Beardsley's view of this is reflected in his lifelong emphasis on the autonomy of the text, a central axiom of his literary aesthetics.

The Principle of Autonomy

In *The Possibility of Criticism*, Beardsley sets forth his 'Principle of Autonomy', the principle that 'literary works are self-sufficient entities, whose properties are decisive in checking interpretations and judgements' (1970: 16). This autonomy is a feature of the linguistic nature of texts – a literary work is 'first of all a text, a piece of language' and, as such, is divorced from both author and critic (17). In 'The Intentional Fallacy' the two critics assert that 'the poem is not the critic's own and not the author's (it is detached from the author at birth and goes about the world beyond his power to intend about it or control it)' (1962: 93). Specifically, the poem is 'embodied in language', 'embodiment' suggesting that it is not itself (identical to) language or is at least separable from language-in-which-it-is-embodied (ibid.). This, Morse Peckham refers to as Wimsatt and Beardsley's doctrine of 'semantic autonomy' (although it is debatable that Wimsatt ever committed to such a doctrine), the result of which is a separation of poetic and ordinary language, the suggestion being 'that poetry has unique semantic functions, different from those of all other kinds of linguistic utterance' (1976: 142). Indeed Wimsatt and Beardsley, in their definition of poetry as 'a feat of style by which a complex of meaning is handled all at once' do contrast it with 'practical messages, which are successful if and only if we correctly infer the intention' (1962: 93). Notice that the distinction is on the grounds of intention. Against this view, the authorial intentionalist traditionally emphasizes literature's communicative or expressive role, playing thus upon the putative similarities between ordinary and poetic language, the latter as a development or special case of the former. As we shall see, an intentionalist view of the literary text (or rather, authorial intentionalism or expressionist intentionalism, I shall later argue that other intentionalisms are possible) presumes an analogical connection between the two uses of language.

Beardsley's characteristic statements on the text's accessibility – it has 'a public character, a determinable quiddity, that lays it open to interpretation' (1970: 68) – and the true focus of interpretation are based in large part upon his Principle of Autonomy, thus: 'We are not judging the poet, or his performance or execution, or his culture or his psychological states, or anything but the poem itself' (ibid.). In this last statement he separates the work from its creator, its inception and its context. More generally, all such statements reinforce the pseudo-scientific tone of his enterprise of treating criticism as a cognitive discipline, where the textual 'knowledge' to be unearthed is to be distinguished from historical or biographical or psychoanalytic knowledge; the critic must decide which sort of knowledge is relevant, mostly in order to distinguish his métier from that of the psychoanalyst or biographer. However, it is arguable that literature resists such treatment – are work and world so easily divorced? – and that such determinability of inside and outside requires a givenness of surface qualities which is, at the very least, suspect.

Hermerén, in his detailing of different versions of the autonomy thesis reads Beardsley's Principle of Autonomy as an approximation of the thesis of autonomous *qualities*: the claim that only the qualities *belonging to* (i.e., internal to) the work of art in question can limit interpretations of it, not the intentions of the author or the responses of the reader. This raises a particular problem: 'The qualities of the work are not given once and for all [. . .]. It is therefore not *always* easy to distinguish between what is in the work of art and what is merely in the imagination of the beholder' (1983: 38). Irony is just such a quality – does it inhere in a text or is it projected upon it? Intentionalists have laboured this point: that language and syntax alone will not reveal ironic intent. Is irony then to be considered as something other than a 'quality' or feature of the work (something which inheres in the relationship between work and reader, perhaps) and therefore something which lies beyond the critical remit of the interpreter? This would be a difficult argument to clinch. The work, says Hermerén, 'is constituted by a dialogue of questions and answers' (ibid.); but, for Beardsley, the work has no such dialogic fluidity. Instead it is objectively given, stable, determinable, present – or must be, if his critical methodology is to convince.

Colin Lyas, though, takes issue with this assumed stability, attacking Wimsatt and Beardsley's 'suspect theory of what it is for something really to be a property of a thing' on the basis of which they aim to exclude all references to artists and authors in criticism (1983: 292). This he labels 'anti-personalism', a stronger thesis than anti-intentionalism, and reminiscent of Lewis and Tillyard's book *The Personal Heresy* (1939) from which Wimsatt and Beardsley took their inspiration. For the New Critic, only those features which can be 'read off' (or from) a work (Wimsatt's phrase in 'Genesis') can be counted as truly 'belonging to' that work, and so relevant to criticism. For example, primary qualities such as colours and sounds and – emerging from these – qualities such as grace, unity and so on, all of these fall into the category of internal qualities or surface qualities. Excluded from this category are qualities such as irony or insincerity, qualities which may require some knowledge of context or authorial intent and which may not be discernible by inspection of surface properties alone. The privileging of surface features is part of Wimsatt and Beardsley's project of excluding from the critical focus all but 'the work itself', a purifying exercise in which surface properties are suggested to be less corrigible than context-dependent ones due, in no small part, to their non-inferential nature. Lyas disputes this assumption of the purity of surface properties, on the grounds that what is less corrigible is not, by necessity, more objective, and on the basis that inference may be involved even in the recognition of surface properties. This is an argument at least partially substantiated by the existence of interpretative disagreement even among avowedly formalist critics[2]; what they are disagreeing about are these supposedly given and incontestable surface qualities.

There is, in addition, an anti-intentionalist argument that those features of a work of art which are not surface features of it are *aesthetically* irrelevant to the consideration of it. However, what makes something aesthetically relevant, that is, relevant to the consideration of the work of art *as work of art*, cannot but be contentious; there are no particular qualities which are predicable only of works of art and not of other objects which might similarly invite aesthetic judgements, such as a sunset or a broken tree bough on the ground. In fact, the distinction between art works and such natural or 'found' objects (or vistas) has historically been made on the basis of the artefactual nature of works of art.

There is no reason to suppose that aesthetic interest in an object *just is* an interest in its surface features. More fundamentally, can we accept in the first place (as Lyas at least appears to do) the problematic definition of so-called surface qualities and the model of surface and depth upon which it is premised that, in turn, reveals a tacit set of assumptions concerning the space and shape of a literary work? The language of surface and depth also implies a kind of aesthetic or interpretative hierarchy, albeit one which Wimsatt and Beardsley overturn in their privileging of the former over the latter, and it is further indicative of the formalist tendency to think of literary texts in spatial terms, to the exclusion of their temporality, their existence in and through time. Even if we do accept such a model, it must be conceded that Wimsatt and Beardsley are hasty in their exclusion of other (allegedly non-superficial) features of a work, a sign of their enslavement to a pre-existing critical ideal. Lyas, by contrast, favours a more open and practice-oriented approach, claiming that it is not possible to discover 'in advance of the study of particular works, a criterion which will tell us what *kinds* of features will or will not be relevant' and that relevance is determined by 'whether in practice a knowledge of that fact affects our appreciation' (1983: 304). However, the business of constructing a critical methodology usually does determine relevance in advance; indeed, the kind of openness in interpretation that Lyas gestures towards perhaps obviates the need for anything as concrete and stipulatory as a critical methodology. Furthermore, as Beardsley and Wimsatt reveal in 'The Affective Fallacy', they are as little interested in the effects of a work of art as they are in its causes.

There is a more general, ontological problem with the Principle of Autonomy which is also the staple, humanist response to Wimsatt and Beardsley: the difficulty and undesirability of dissociating the text utterly from author and reader (or, more generally, from the realm of human endeavour). As Cioffi says, 'there is an implicit biographical reference in our response to literature. It is [. . .] part of our concept of literature', adding later that 'the suspicion that a poetic effect is accidental is fatal to the enjoyment which literature characteristically offers' (1976: 66, 68). The pleasure of reading, then, is founded upon some idea of the human effort involved in the creation of that 'poetic effect', an effort which – rightly or wrongly – we interpret as less random and more meaningful than a 'mere' effect of language, words happening to rub up against each other in an interesting way; this question of why the intentionality of authors is traditionally perceived as so much more compelling than the (arguable) intentionality of language can perhaps be returned to at a later point. Sparshott similarly disputes the separation of performer and performance that the Principle of Autonomy entails, arguing that the artist is more than 'a bystander at his own performance' and that the separation 'depends on our taking the performance as an aesthetic object rather than a work of art, as if it were a natural object and not a performance at all' (1976: 107, 108).

The distinction, then, although this is not explicitly stated here, is one between 'natural' objects and intentional objects, the latter being either the outcome of or even isomorphic with some kind of 'performance'. Beardsley does indeed regard the work as an 'aesthetic object' rather than a 'performance' and it is on this basis that he argues the irrelevance of authorial intentions and the distinction of object from intention because 'the evidence for the existence and nature of one cannot be exactly the same as the evidence for the existence and nature of the other' (this is an argument from logic, owing something to Leibniz's Law on the identity of indiscernibles) (Beardsley 1958: 19). In fact, Beardsley seeks to separate object and intention on the grounds of their different intrinsic properties, to show that they cannot be identical, or even directly related. He claims that the intention gives us only 'indirect evidence' of the aesthetic object, and vice versa (20).

Only an (ontological) account of literary work as performance can bring work and intention together, but this will then be subject to the same objections historically levelled at other instances of expressionism: predominantly the accusation of oversimplification in postulating a single guiding emotion which fuels the creative process, or a single meaning to the communication of which the work is devoted. The combined implications here of the weakness of a monistic authorial intentionalism raise the question of why the intentionality of an artwork must always be linked to the author – not only to consciousness, but to his consciousness in particular. Such privileging of the author rests on the kind of humanistic approach to literature which the New Critics (among them Wimsatt and Beardsley), and later those working under the banner of structuralism, were attempting to replace with something more formal, structural, something 'purer'. If the humanist's thesis that 'every work of art is a human achievement' is sidelined, so too is the author: first, in practice and, latterly, as an organizing, dogmatic concept (Jones 1964: 142).

The role of the author

Intentionalism as figured (and opposed) by Wimsatt and Beardsley just is authorial intentionalism. They – and their anti-intentionalist counterparts and successors – rarely qualify it in this way, although it is precisely this qualification which I take to be intentionalism's weakest point, inviting attacks on psychologistic grounds (the author's intentions, thoughts, motivations, etc., are not accessible to the critic) or merely on the grounds of relevance (the object of criticism is the work, not its author). There is the further supposition that intentionalism is committed to a view of the author as ideal reader of his works, and that authors' statements on the meanings of (or intentions behind) their own works cannot, therefore, be contradicted. In fact, few intentionalists would claim that the author has a special vantage point from which to view his work, excepting the obvious cases of neologism or syntax peculiar to that author; to credit the role of the author's intention is not to believe everything that the author subsequently tells you about the work. Indeed, Isobel Hungerland proposes that the author's statements concerning his work are of historical or biographical (rather than critical) interest only and, moreover, that:

> The author's intention is [. . .] a standard only in an odd sense of 'standard'. Approximation to the author's reading is the criterion of my choice just because my purpose is to achieve this. For a man with another purpose, the choice may be reasonably made on another basis. (1955: 738–39)

Thus she goes some way towards reinforcing Wimsatt and Beardsley's argument. However, others (e.g., Kemp 1964, Lyas 1983, Jones 1964) have suggested that the author's word has some advisory force, and may lead the reader or critic to see something within the work of which they had not previously been aware, or similarly may lead them to eliminate a particular reading as inappropriate. If corroborating evidence exists – if it is 'objectively there' within *the work itself* (Kemp 1964: 151) – then this may be said to contravene the Wimsatt and Beardsley ban on external evidence in practice perhaps, but not in principle, for the deciding factor is still the work.

Wellek and Warren are of the camp which warns that *expressions* of intention, as distinct from intentions themselves, may not be reliable. Such expressions are always 'rationalizations' and must be 'criticized in the light of the finished work of art' (1963: 148); there is little to disagree with here. Like Wimsatt and Beardsley, though, they cling to a blueprint account of intention and are therefore keen to emphasize the gap between (prior) intention and actual achievement. The possibility of artistic failure is the weapon with which the intentionalist is all too frequently beaten. Kemp picks Mrs Malaprop as his paradigm failed communicator to bolster an argument that it is not what the author (or, in this case, speaker) intends to mean that matters, but what he *succeeds* in meaning. In fact this is a spectacularly bad example: Mrs Malaprop is, by design, a humorous character, and the humour arises *only* from knowledge of what she *intends* to convey and the extent to which her actual words misrepresent and deviate from this intention; if we had no (inferred) knowledge of intention here (and there are many textual clues) the words themselves would be nonsensical, even meaningless. It is notable, however, that for clues to Mrs Malaprop's intention and to that of her creator's, Sheridan, we need not step outside the work itself or make the mistake of 'consulting the oracle' (Wimsatt and Beardsley 1962: 104).

The suggestion that the author may not be the best judge of his own performance is a question of evaluation and a separate matter; only the most suggestible critics would allow an author's own judgement on his achievement to colour their own. In 'The Intentional Fallacy' it is evaluation with which Wimsatt and Beardsley are primarily concerned, but Wimsatt later manipulates their original thesis into the enlarged claim that: 'The intention of a literary artist *qua* intention is neither a valid ground for arguing the presence of a quality or a meaning in a given instance of his literary work nor a valid criterion for judging the value of that work' (1968: 12). There are two interesting developments here: first, the idea of 'intention *qua* intention', which implies a refinement of the original, limited, blueprint conception; second, the inclusion of questions of meaning and interpretation which, Wimsatt claims, they had meant to be integral to their thesis all along; an interesting, after-the-fact rationalization! Perhaps, though, this is a recognition of the fact that intentionalism in matters of evaluation was hardly worth contesting to begin with, unless they were taking issue with a view such as that promulgated by Hungerland, namely, 'we cannot appraise the skill of a performer or producer without knowing what his objectives are' (1955: 740); 'objectives', though, are easily dismissed by dint of their susceptibility to change. Also, although knowledge of an author's intentions and objectives *may* alter our view of his skilfulness in carrying them out (e.g., if it is revealed that a piece we thought fatuous or simplistic is in fact intended as pastiche or parody), yet it is perfectly possible to judge an author's skill without reference to such objectives. *Statements* of such objectives may be more or less honest, more or less sincerely motivated, but such honesty or sincerity is hardly the concern of the literary critic.

T. M. Gang argues that it is only in cases of ambiguity or difficulty in comprehension that it makes sense to ask after the speaker or writer's 'literary' intention (i.e., to ask what he meant to signify, what meaning the words were intended to convey). Gang contrasts 'literary' intentions with 'practical' intentions, the latter aim at a particular end or result and are manifest in behaviour although, as we will see, Wittgenstein's anti-mentalistic characterization of intention as something which *just is* manifest in behaviour may be read as undermining this distinction. If something is 'quite plainly and unambiguously said' then the words uttered or written display the intention better than any paraphrase of them, which is all that a statement of intention would then amount to (Gang 1957: 178). This last point presents an irresistible paradox: the writer can best express what he meant by the misunderstood or ambiguous line of text by repeating that line; to do otherwise is to concede an erroneous or ill-advised choice of words or syntax in the original version. This, too, has been employed as an argument against the futility, the counter-intuitiveness of intentionalism, but it is only *authorial* intentionalism against which it can be so effectively wielded.

Wimsatt and Beardsley contend that 'the meaning of words is the history of words', that is, the history of their usage (1962: 98). It is usage which confers meaning, and usage is figured as a supra-personal, supra-historical force. Therefore, the author is always in thrall to this greater power: language and the history of linguistic usage, along with conventions which cannot be overridden if he is to make himself understood. The poet uses words 'and they have a meaning independently of his having used them', although he may yet 'modify their meanings by using them in the way he does' (Roma 1976: 81–82). So, a writer cannot 'intend a meaning into' a word. Nevertheless, he does *use* that word *intentionally* and, presumably, in full knowledge of the history and conventions of its usage, and the ramifications and implications of its usage in this context. This is intention at the most basic level or, if you prefer, a kind of metaphysical tussle between intention and convention, both with contributions to make. Roma and Gang (among others) overlook this, implying that it is only in the case of neologisms or usage contrary to convention that intention plays a role; this is a stronger claim than the thesis that it is only in such instances that intentions should be *referred to.*

The privileging of the author is frequently seen as a determinedly *ethical* imperative springing from the intuition that the poem *belongs* to its creator in a manner distinct from the way in which it is subsequently appropriated by its reader(s) and critic(s). Therefore, as readers we have a responsibility to the author, we feel compelled to read his text in the way he would have wanted. Many intentionalist arguments play upon this moral compulsion of the reader to respect authorial wishes (and intentions); as will be considered in the next chapter, E. D. Hirsch offers a kind of 'ethical' argument (the 'need' for a normative criterion in interpretation and the eminent suitability of authorial intent for this role) which faces objections on logical grounds. In fact most ethically inclined intentionalisms struggle to delineate the nature of this 'belonging', the implied sacrosanct relationship between author and work; a 'belonging' which goes beyond seeing the author as 'cause' of the finished work, and incorporates also some of the institutionalized reverence with which we regard artists. There is an element of deliberate mystification in this, entailing as it does a view of the creative process as occult or, at least, ungraspable by the uninitiated. Nevertheless, there is the practical relevance of the author as instigator or performer of a communicative act: it makes sense to ask not only what is being communicated but also who is doing the communicating, and this may justifiably colour our understanding of what is 'said'. How do we mediate between these two positions: the ethical, which may be seen as undemocratic or elitist or exclusionary, and the permissive or pluralist position which may encourage a certain critical anarchy or amorality? Of course, criticism (or, more simply, *reading*) may be seen as an essentially amoral activity, in which the usual ethical concerns are somehow suspended or in which our responsibilities are directed towards an object (the text) rather than towards a person, a redirection which at least entails a lowering of the stakes (this is art, not war). There remains, though, the practical problem of juggling our responsibilities (wherever they are directed) without succumbing either to occultism or relativism.

Author's intentions, speaker's intentions

In talking of the author – the designing, deliberating consciousness – whether for or against, the participants in the intentionalist debate make frequent reference to the 'speaker', that is, the narrator of the novel or the voice of the poem. This raises the problem of whether the author shares the beliefs of his speaker and wishes to assert those propositions asserted by this speaker. This is a question of the different strata of intentions to be found within a literary text and the subtleties are often overlooked in an easy conflation of author and speaker. In 'The Intentional Fallacy', Wimsatt and Beardsley state that 'We ought to impute the thoughts and attitudes of the poem immediately to the dramatic *speaker*, and if to the author at all, only by an act of biographical inference' (1962: 93). They are correct to distinguish between author and speaker, but not to see the latter as free of the jurisdiction of the former. Jurisdiction, however, is not conflation although Beardsley appears to believe that intentionalism entails just such a conflation, arguing in *Aesthetics* that the (intentionalist) claim that 'what the *poem* means' is discoverable via 'what the poet meant' implies, 'that the dramatic speaker, the "I" in the poem, is always the author of the poem, so that any evidence about the nature of either of them is automatically evidence about the other' (1958: 24). He is wrong to impute such a view to intentionalism (and I can find no evidence that there exists among intentionalists an inability to distinguish between an author and his narrators): in fact, the implication of the intentionalist claim is the weaker thesis that the speaker of the poem displays the attitudes and beliefs that the author *intends him* (the speaker) to have, not necessarily the attitudes and beliefs that the author either holds himself or wishes to endorse. After all, the author is writing not autobiography but *fiction* and even in writing which presents itself as non-fictional a narrative position or character or voice is adopted which does not necessarily coincide with that of the author; while he may intend to *represent* certain views, it is a false inference to claim that such views are therefore his own. The text is, if anything, the *effect* or product of the author's desires and so *distinct* from them, as any effect is distinct from – *although not unrelated to* – its cause. Beardsley appears to have missed something crucial in the nature of what it is to write fiction, to write representationally – whether of events or attitudes. In doing so he has also brought to light the troublesome logical status of propositions within fictional texts, the question of whether they constitute actual assertions or something possessed of less force.

In his 1982 article 'Intentions and Interpretations: A Fallacy Revived', Beardsley consolidates his earlier mistake, but here using the terminology of 'illocutionary actions'. He interprets 'illocutionary actions performed intentionally' as 'illocutionary actions performed with the intention of performing that action' (196), when plainly this is not the case in fictions. To reiterate: if the writer represents a promise (a kind of illocutionary action), he does not himself intend to promise, but he does intend to represent (or perhaps 'simulate') a promise. The altered conditions of meaning and reference do not divest the action of representation of its intentionality. It is Beardsley here who is failing to distinguish between author and speaker.

Of course, the sharp distinction made between author and speaker in 'The Intentional Fallacy' is commensurate with Wimsatt and Beardsley's project of giving boundaries to the work (or, rather, boundaries to what is acceptable in criticism) and thereby forbidding reference to all that is 'external'. The speaker is internal, a feature of 'the work itself'. It is increasingly evident, then, that it is the author, not intention per se (or even intention *qua* intention), that they are attacking. They are, properly, anti-personalists in the tradition of Lewis and Tillyard, not anti-intentionalists at all. The autonomous text, the 'embodied' language of Wimsatt and Beardsley has no need for the author; criticism is not biography or psychoanalysis; human concerns are not artistic concerns, the former are aesthetically irrelevant. However, this does not mean that this uniquely autonomous entity which has 'a will, or at least a way, of its own' is characteristically *non-intentional* (Beardsley 1970: 37); as we have seen, 'The Intentional Fallacy' does not preclude discussion of intention-in-the-text.

Interestingly, Wellek and Warren concede that 'there can be no objections against the study of "intention", if we mean by it merely a study of the integral work of art directed towards the total meaning.' (1963: 149) This is a definition of intention suggestive of a teleological impetus or ordering within the work itself; perhaps even an *impersonal* definition or attribution of intention. However, the lines receive the disclaimer that 'this use of the term "intention" is different and somewhat misleading' and Wellek and Warren do not proceed any further in their analysis of the possible intentionality of the 'integral work of art' (ibid.). Nevertheless, Roskill, unusually, does examine the ways in which the term 'intention' might be predicated of the work rather than the author, and charts the history of its usage – something that few in either the intentionalist or anti-intentionalist camp bother to do. He claims that there is 'no real Latin equivalent for the term "intention"' (1977: 99). Instead, *voluntas, animus, consilium* refer to mental states, or some conjunctive phrase is used to denote purpose or aim. *Intentio* implies effort or agency, and in Medieval Latin it is predicated equally of people and things. By the nineteenth century, however, Roskill notes 'the beginning of that transposition to the third person', which is relevant to his discussion of the 'object-oriented use of "intention"' and as far as texts are concerned, intention now refers to 'an underlying scheme of thought or ideas which is to be unfolded' – a construction comparable to Wellek and Warren's tentative fumblings in that direction (Roskill 1977: 100). This concept of artistic intentions (intentions belonging to or predicated of the work), as contrasted with artist's intentions, has contextualist and generic variants – thus we may talk of the artistic intentions of a particular period or a particular style. Ideas of teleology or orientation reappear again and again, also ideas of the work's function. In the history of art, the concept of *kunstwollen* (artistic intentions) stems from the existence of the collective of artists in which the individual artist is a mere (un-individuated) part, his own personal intentions unrecorded; there is precedent, then, for the 'third-personal' use of 'intention', and it is this idea that will be pursued in the pages that follow, via a search for intimations of such a usage of 'intention' in more recent theorizations of the literary. To predicate intention of an *object* (in this case, a literary text) is, by necessity, to distance it from the mental and from consciousness.

Demystifying the mental

A. J. Close writes that 'a major source of confusion is to see intention as a "mental objective", connected with action contingently or causally but not logically' (1976: 176). This is to equate 'intending' with Aristotle's 'formal cause', that is, to see it as 'a prefixed, static, mental norm by which the artisan's physical work (Aristotle's "efficient cause") is guided, and from which it may well deviate' (ibid.). The mentalistic conception of intention ignores the practical aspect of intention involved in doing or acting. Intention cannot be separated from action – 'the practical knowledge which intentional action involves is essentially displayed in *doing*' (ibid.). Such a separation raises several problems: if intention and action are only contingently related, how is the latter 'controlled' by the former? Also, we are presented with the difficulty of how to explain the absence in many instances of 'a mental, normative picture of one's completed action' (Close 1976: 178), for example, in those instances of spontaneous, yet intentional, behaviour, when we do not meditate upon a desired outcome of any sort.

Wimsatt and Beardsley's anti-intentionalism owes much to a Cartesian account of mind and body as contingently related and, therefore, separable; by analogy they treat author and text as mind (private, inaccessible) and body (public, observable). The flaws of the dualist account of mind and body have been well documented. Gilbert Ryle writes, famously, of the 'official theory' which he titles 'the dogma of the Ghost in the Machine', noting that:

> When someone is described as knowing, believing or guessing something, as hoping, dreading, intending or shirking something, as designing this or being amused at that, these verbs are supposed to denote the occurrence of specific modifications in his (to us) occult stream of consciousness. Only his own privileged access to this stream in direct awareness and introspection could provide authentic testimony that these mental-conduct verbs were correctly or incorrectly applied. (1949: 17)

The project of Ryle's *The Concept of Mind* is to question the counter-intuitive yet hitherto rarely disputed notion that, 'a person [. . .] lives through two collateral histories, one consisting of what happens in and to his body, the other consisting of what happens in and to his mind' (1949: 11).

Wittgenstein's extemporizing on the knowledge of other minds also works to contradict the Cartesian account, simultaneously paralyzing psychologistic objections to intentionalism:

> If I see someone writhing in pain with evident cause I do not think: all the same his feelings are hidden from me [. . .]. It is possible to imagine a guessing of intentions like the guessing of thoughts, but also a guessing of what someone is actually *going to do*. To say 'he alone can know what he intends' is nonsense: to say 'he alone can know what he will do', wrong. (1953: II, xi)

In other words, what is internal is not necessarily hidden from us – 'I can be as *certain* of someone else's sensations as of any fact' (ibid.). Beardsley's conception of intention, however, is inextricably linked to awareness and the mental, as is evident in his 1978 article 'Intending'. Here he distinguishes *doing intentionally* from *intending to do* and *having an intention*, and notes also that conflicting wants do not affect the definition of doing something intentionally. He gives a '"know-and-want account" of intentional doing' (1978b: 166) which stipulates that if an action is to be categorized as intentional, the agent must have the appropriate awareness and 'yen' to do the action. This is not as strong as Fleming's *causal* know-and-want account (1964) – later refined by Goldman (1970) – which posits that to perform an action intentionally, the agent must: know he is doing it, want to do it and do it *because* he wants to (i.e., not by accident); but neither does the causal account contradict Beardsley's. Of prime importance to Beardsley's account is the agent's possession of the requisite belief state – to intend to do an action, one must believe that one can and will do that action – and this leads him to define intention as 'a co-referring want–belief pair' (1978b: 180). Thus, despite the connection with action, it is the (supposed) mental origins of intention that mark it out for what it is. He rather conveniently bypasses the issue of other minds, ignoring the fact that we every day apprise and recognize and act in response to the intentions of others, we 'read their minds' with scant confusion, scant mystification.

However, despite the compelling evidence of experience, critical heritage dictates that intentions will be easy to dismiss as long as they are associated with the mental states of the author. This leads me to the conclusion that the way forward must be to follow Roskill's example in locating intention in (or predicating it of) 'the work itself' – or even in language itself. This is a bolder project, but not therefore unattainable.

The way forward (a brief polemical conclusion)

A close reading of 'The Intentional Fallacy' has shown that there are degrees of anti-intentionalism and that if Wimsatt and Beardsley are guilty of anti-intentionalism it is to no great degree. In fact, their motivation appears to be methodological not metaphysical and their exclusion of the author – as a complicating factor, contaminating a more 'pure' structural approach to criticism – does not in itself involve an exclusion of intention. 'The Intentional Fallacy' could more accurately be titled 'The Biographical Fallacy' or 'The Genetic Fallacy'. Furthermore, both writers' later statements on the matter suggest room for manoeuvre, and both concede that their original formulation of intention was a narrow one. On the scale of anti-intentionalism plotted by Lyas (1973: 197), there are three versions of the anti-intentionalist thesis, in ascending order of strength: first, ban all references to *statements* of prior intention; second, ban all references to *prior* intentions; third, ban *all* references to intention. Wimsatt and Beardsley never adopt a position as radical as the third. Their legacy, however, lies in the fact that their descendants have failed to realize the mildness of their original thesis, and it is this subsequent theoretical work that needs to be undone – or at least substantially interrogated, its erroneous premises exposed.

Profiting from this reading of Wimsatt and Beardsley, a more acceptable version of intentionalism presents itself: one which figures intention as internal to the text, distanced from consciousness, accessible, part of the text's structural make-up. A. J. Close writes that:

> The intentions that we assess [in literature] [. . .] are simply an intrinsic feature of linguistic acts – i.e. the meaning of acts of communication – and not a matter of well- or ill-intentioned consequence which [. . .] we tend conceptually to separate from its action. (1976: 177)

My project, then, is to explore, delineate and develop this idea of intention as 'an intrinsic feature of linguistic acts' with special application to the understanding of literary texts and so to re-introduce this troublesome concept of 'intention' into the literary debate from which it has been excluded for so long. This will involve, in no particular order: a full discussion (with borrowings from philosophy and linguistic theory) of the intentionality of language, and thus the fundamental inextricability of meaning and intention; an examination of the extent to which texts can be viewed as 'linguistic acts' or utterances – and the extent to which this is a vital under-determination of them; and an encounter with deconstructive theory – as the apparent exponent of a late-twentieth-century variant of anti-intentionalism which yet leaves space for a consideration of a more generalized formal intentionality. These separate but interlinked analyses will decide the strength or poverty of intention so figured. Prior to that undertaking, however, I propose to start by looking at the leading intentionalist responses to Wimsatt and Beardsley – responses normative (Hirsch) and logical (Juhl) – and at some more recent meanderings in intentionalism. It need not be concluded, because all versions of intentionalism have hitherto failed to convince, that intention itself (intention *qua* intention etc.) must become a concept similarly jaded, similarly inert and, finally, dispensable. Let us consider what we are throwing away.

Notes

1 In fact there are myriad applications: anterior (causal) and ulterior intentions, intentional and voluntary actions, counter-intentional and involuntary actions, intentions *in* acting, purposive and non-purposive intentions and so on.

2 See, e.g., the radically opposed interpretations of 'A slumber did my spirit seal' given by Cleanth Brooks and F. W. Bateson, which Hirsch cites in *Validity in Interpretation* (1967: 228).

Chapter 1

'The Soul of Speech': E. D. Hirsch and the Ethics of Authorial Intentionalism

The ethics of autonomy

What does it mean for an interpretation to be 'ethical'? Can it be ethical in effect if not in fact ethically motivated? What is the connection between the ethical and 'the human' or humanism – does post-structuralism (as a variety of anti-humanism) have any use for ethics? Can both monistic and pluralistic theories be ethical (that is to say, to what extent is semantic determinacy a feature)? These are some of the questions which arise in seeking a definition of ethical criticism. I will not attempt such a definition here but will be considering the ways in which E.D. Hirsch's particular brand of intentionalism – his riposte to the New Critics – can be described as 'ethical' and whether this description buttresses his position or lays him open to accusations of arbitrariness. Either way, Hirsch's ethical tone will be seen both to capture the essence of his defiantly anti-ontological argument and to reveal a contradiction in his view of criticism as (ideally) analytical, pseudo-scientific (fact-based) and yet somehow irrevocably tied to the human (value-based). The difficulty lies in his attempt to formalize and give substantive methodological grounds for ideas (such as the centrality of the author) which seem to him to be *pre-theoretically* convincing or valuable.

However, it is not only intentionalism that may occupy the moral high ground, according to Tobin Siebers, who sees a covert ethical agenda behind the semantic autonomy of Wimsatt and Beardsley. In their removal of the author from the scene of interpretation, says Siebers, the New Critics create a separation between poet and poem which 'acts to shield authors from aggressive criticisms' (1988: 48). The 'work itself', and perhaps its dramatic speaker(s), carry the responsibility. However, Siebers does not consider the ethical ramifications of absolving the *critic* of his responsibility in this way: what kind of ethical imperative, if any, does responsibility to the text, an apparently inanimate, unresponsive object, constitute? How impotent an imperative is it once the text is divorced from its social (human) context? Siebers implies that social responsibility is not an issue as far as fictional texts are concerned, but legal texts are another matter, thus 'the New Critics simply cannot dispose of intention in the case of murder because social responsibility is at issue', but they can do so in the case of poetry, where it is not (1988: 54). Here Siebers is drawing a line between poetic autonomy and an all-out textual autonomy which would deny the socio-political ramifications of certain non-fictional texts. Like Wimsatt and Beardsley he relies upon an assumed clarity of distinction between poetic and 'ordinary' language or, just as nebulously, between the ways in which language is *used* in literary and non-literary situations, which is at least contentious.

The most significant flaw in Siebers' 'ethical' reading of 'The Intentional Fallacy' is his failure to distinguish between, on the one hand, the unintentionally ethical *effect* of a theory (which this 'protection' of the author is, and it is nowhere clear that Wimsatt and Beardsley are even aware of this possible effect of their critical method); and on the other hand, intentionally ethical *motivations* (which are not in evidence here). In any case, Wimsatt and Beardsley's shielding of the author can only be considered 'ethical' if 'the author', whether historically situated individual or classificatory construct, is considered worth shielding; of this, more later. In my previous discussion of 'The Intentional Fallacy', I concluded that Wimsatt and Beardsley are motivated primarily by a desire to streamline the discipline of criticism and to create a manifesto of formalism; in other words, their main preoccupations are aesthetic, methodological. But an aesthetic or methodological ideal – to purge critical writing of its biographical superfluities – is not an ethical ideal. Siebers sees the anti-pluralism of Wimsatt and Beardsley as ethical by definition and in doing this he is proposing a kind of ethics of the text, seeing poetic autonomy as a worthy successor to the human autonomy of Kant. Would he be as happy with any interpretative monism, however arbitrary, provided it succeeded in fending off relativism, and allowed the critical enterprise a determinate goal at which to take aim?

The move away from a focus on the human qualities of the text (figured as an expression of human subjectivity) and its humane effects (in allowing a readerly identification with and understanding of such an expression) to a focus on language, structure and the text's formal qualities, is more marked in the critical movements which succeeded New Criticism. Abandoning the methodological monism of Wimsatt and Beardsley (which reassures the critic of possible-in-principle semantic determinacy), post-structuralists have yet retained the semantic, structural emphasis on the 'text itself', while substantially expanding upon what exactly is meant by 'text'. Contra Siebers, their pluralism can also be seen as ethically motivated (gaining force from the radical politics of the late 1960s from which it sprang) if conceived as the refutation of totalizing gestures in criticism and the embracing of the socially and/or politically marginalized or dispossessed, via an analysis of the marginal and meandering meanings of the text.

More will be said about the tenability of ethical analysis with regard to these later phases of formalism at an appropriate stage in the discussion. At present it is sufficient to note the contemporary association of critical judgement (or any judgement) with power and violence which Siebers fears could lead to the end of the practice of criticism, witness his avowal that 'literary criticism cannot endure without the freedom to make judgements and modern theory urgently needs to regain the capacity to decide' (1988: 41). The result of this alleged fear of decision-making with its totalitarian overtones has been the privileging of contingency, heterogeneity, irrationality: qualities which literary texts are held to possess in abundance, thanks to their polysemous nature and their openness to re-reading and re-interpretation.

Siebers claims that it is the New Critics who first recognize and embrace 'literature's chaotic nature as representative of its autonomy' but for Wimsatt and Beardsley the text is still a stable object of observation and consideration, the meaning and quality of which may be determined, evaluated and judged, its 'chaotic' nature and its 'autonomy' notwithstanding (1988: 34–35). The critic is not yet in thrall to the text, but neither is he in thrall to the 'oracle' author. By contrast, Hillis Miller's subsequent 'ethics of reading' stipulates that the reader must 'allow the paradoxical and undecidable character of textuality to shape and coerce them' (Siebers 1988: 38). The shifting emphasis from the human to the linguistic has not meant an abandonment of ethics for Hillis Miller, but it has amounted to a fundamental change in what we think of as 'ethical'. No longer defined according to its relations to human subjectivity or social context, Hillis Miller's 'ethics' of reading is defined as 'the moral necessity to submit in one way or another, whatever one says, to the truth of this linguistic imperative', that is, to the undeniable indeterminacy of language and its coercive force (Konigsberg 1981: 41). The text is no longer merely autonomous, now it has a kind of interruptive or interpolative power, as far as our judgements and opinions of it are concerned, a development that Wimsatt and Beardsley could never have predicted. This is a question of agency and, therefore, of intention (the inextricability of the two concepts will be touched upon in the next chapter). It is a question of whose intention counts in the construction of meaning, and the gradual – but decisive – re-characterization of agency as a formal or textual rather than a specifically human quality is the story of the slow burial of the concept of intention. My suggestion is that intention will only survive as a workable concept if it is similarly rethought and recast.

'The soul of speech'

E. D. Hirsch, however, indulges in no such recasting. His argument for the primacy of authorial intention rests on a conviction of the ethical supremacy of this position given the critic's responsibility to, in no particular order of preference, the author and the discipline of criticism. For Hirsch the necessity of decision (regarding the choice of a normative criterion for interpretation) is not only non-violent, it is a determinately *ethical* obligation. It is the centrality of free choice that makes his position ethical, choice that he argues is unconstrained by logical imperatives arising from the nature of the text (or the nature of language). He hints at this in *Validity in Interpretation* (1967) in claiming that 'the object of interpretation is no automatic given, but a task that the interpreter sets himself. *He* decides what he wants to actualize and what purpose his actualization should achieve' (25).

Furthermore, while he argues against semantic autonomy on the grounds that 'no logical necessity compels a critic to banish an author in order to analyze his text', he does not counter this with a *logical* (or ontological) argument of his own for the relevance of authorial intention as an interpretative criterion (1967: 2). The tacit assumption of an ethical stance in *Validity*, most evident in Hirsch's puritanical tone (and hyperbolic tendency) when criticizing the 'wilful arbitrariness and extravagance' encouraged by exegetical criticism, metamorphoses into a declared interest in ethical persuasion in *The Aims of Interpretation* (ibid.). Here he argues that: 'The choice of an interpretive norm is not required by the "nature of the text", but, being a choice, belongs to the domain of ethics rather than the domain of ontology' (1976: 7).

It is in *Aims* that he makes his most explicit statements of the responsibility owed to the author by the critic, the responsibility to accurately determine the author's intended meaning (while also, paradoxically, beginning to employ a conception of meaning less tied to the author, a more impersonal construction which highlights the role of the reader). This is now characterized as a moral injunction but it is also – despite Hirsch's claims to the contrary – a kind of ontological argument as it plays upon the artefactual nature of texts, treating all speech as 'an extension and expression of men in the social domain' (1976: 90). Indeed, his argument is *only* ethical *because* of certain things that he takes to be ontologically given and unshakeable: that texts are uniquely human creations, that they stand in a particular privileged relation to their creators, who are identifiable as such. If we, as readers, did *not* hold such beliefs, although we *might* choose the author's intention as an interpretative criterion, we would not do so for ethical reasons; the ethical argument emerges from the ontological beliefs. This is illustrative, also, of Hirsch's analogous treatment of poetic and 'ordinary' language; he rejects semantic autonomy on the grounds that all language is expression and therefore cannot be divorced from the human (from the particular human doing the expressing). The abuse and misuse of texts (specifically, exegetical 'extravagance' or wilful misapprehension) is as morally reprehensible as the abuse of any person; the moral consequences (if not, perhaps, the practical consequences) of such abuse are on a par:

> When we simply use an author's words for our own purposes without respecting his intention, we transgress what Charles Stevenson in another context called 'the ethics of language', just as we transgress ethical norms when we use another person merely for our own ends. (90)

In this way, and eliding the fact that Stevenson's statement was not concerned with literary texts, Hirsch transfers Kant's moral imperative that men be treated 'as ends in themselves' to 'the words of men' which are not only made meaningful but are themselves 'humanized' (anthropomorphized) by their origination in some creative intention, such that: 'When we fail to conjoin a man's intentions to his words we lose the soul of speech, which is to convey meaning and to understand what is intended to be conveyed' (ibid.).

Without this 'soul of speech' words are 'mute' and even, as Knapp and Michaels will at a later date explicitly argue, entirely meaningless.[1] It is this 'soul of speech', says Hirsch, which compels us to choose authorial intention as our normative criterion; the author has a special claim to the meaning and, therefore, to the subsequent use of his words; furthermore, as his words are an expression of him and so part of him, consulting the author is not, contra Wimsatt and Beardsley, going 'outside' the text. However, while the original Kantian injunction is overtly concerned with personal freedom, multiple interpretations of a work do not directly impinge upon the freedom of its author, however much they may be seen as a distortion of his original vision for that work. In fact, says J. W. Meiland, multiple interpretations do not constitute misreadings as long as they are supported by textual evidence; they are predicated of the work and so need not attribute anything at all to the author (1978: 45). The weaker ethical argument that adherence to authorial intention forces an encounter with the author (i.e., with another person, with that person's arguably unique experience or perspective) which is potentially ethically improving is easily dismissible as encountering another critic's interpretation may prove similarly beneficial and enlightening.

Hirsch himself seems to doubt the tenability of his ethical claims. He therefore offers another argument for intentionalism based on quite different premises: first the belief that the longevity of the discipline of criticism is assured only by its transformation into (or consolidation as) a 'cognitive' or pseudo-scientific enterprise; and second, as mentioned earlier, his own assertion that there is no normative criterion for interpretation structurally implicit in language itself – witness Hirsch's declaration, of great foundational significance to the central thesis of *Validity*, that 'a word sequence means nothing in particular until somebody either means something by it or understands something from it' (1967: 4) – and therefore the critic must choose his own. For this reason, in *Validity*, Hirsch chooses a defence of intentionalism 'that appeals not to the ethics of language but to the logical consequences that follow from the act of public interpretation' (26). He adds further, lest we be in any doubt, that:

> My case rests not on the powerful moral arguments for re-cognitive interpretation, but on the fact that it is the only kind of interpretation with a determinate object, and thus the only kind that can lay claim to validity in any straightforward and practicable sense of that term. (26–27)

The 'logical consequences' of interpretation demand that, for criticism to be a worthwhile discipline, it must be able to claim 'validity': it must be normative.

Hirsch then concludes that authorial intention is 'the only compelling normative principle' (26). Notably he does not claim here that it is compelling on ethical grounds but because it suits his personal goal of validity. There is a suspiciously close link forged here between intentionalism and validity: the former is chosen because it (allegedly) promises validity in a way that other modes of criticism do not, and this goal of validity is presented (as admirable in itself) in order to prove the desirability of intentionalism.

In the previously discussed ethical (and tacitly ontological) argument, Hirsch has asserted that the 'somebody' whose meaning we should adhere to is the author, because of the privileged relation in which the author stands to the text, but in the later 'only compelling normative principle' argument he is conceding (in the choice of 'compelling' over 'possible' or 'logical') that such a principle is arbitrary and must arise out of a sense of responsibility to the 'vocation' of criticism. He himself agrees that it is 'not the only possible norm for interpretation', although he gives public consensus short shrift as a possible alternative, despite his later concession that interpretative disagreements often reside 'in choice of emphasis rather than choice of meaning', which suggests that consensus is not only possible, but likely (1976: 7, 89). Nevertheless, he insists that 'on purely *practical* grounds [. . .] it is *preferable* to agree that the meaning of a text is the author's meaning' (1967: 25; my emphasis).

What should be noted, then, is that the 'practicality' of the criterion is relative to Hirsch's particular purpose (the establishment of a cognitive criticism) and such a criterion is 'preferable' only if it is agreed that cognitive criticism is more desirable (or more ethical, perhaps) than, say, pluralism. Yet there is nothing intrinsically ethical about the goal of semantic determinacy and, as has already been suggested, monistic approaches to interpretation may in fact be regarded as exclusionary or even totalitarian in the way that they insist upon a distinction between 'correct' and 'incorrect' readings, positing the possibility of a 'right answer' in interpretation. Hirsch uncritically assumes the methodological superiority of the sciences and the pseudo-scientific (which alleged superiority Derrida, e.g., dismisses as a fallacy of logocentrism). Although the two arguments offered by Hirsch may more favourably be viewed as complementary – the ethical intuition of a responsibility to the author which simultaneously provides a compelling criterion for objective interpretations – they actually represent two different forms of moral theory, deontological and consequentialist, respectively: the former focusing on duties or obligations, with ethical value lying in right action; the latter focusing on consequences, with ethical value lying in good states of affairs. There is, of course, some cross-classification of these two types of moral theories as the 'right action' is likely to be the action which brings about the best state of affairs (although this will not always be the case), but Hirsch appears to offer parallel rather than mutually sustaining arguments. This is indicative of an attempt to fight the battle against anti-intentionalism on both ethical and epistemological fronts. In *Aims* he divides his central thesis into two claims: one, the reader should try to reconstruct authorial meaning (ethical); two, accurate reconstruction is, in principle, possible (epistemological). His two-pronged attack may then be seen as representative of his belief in 'the correlations that exist between knowledge and value, not just in interpretation but in the humanities generally' (1976: 12).

Even if the more sympathetic view is taken that the ethical and epistemological/ontological theses are successfully interdependent (a point which Hirsch doesn't explicitly argue), the basis of the whole is this ethical commitment to the privileged status of the author, which is subject to the objections levelled at ethical arguments in general: the difficulty/impossibility of rational agreement in ethical matters, and the questionable semantic status of moral judgements – whether they are (mere, subjective) prescriptions rather than (indubitable) statements. Hirsch appeals to our moral intuitions, as well as to more general intuitions about what kind of thing a literary text is, but this approach means that his thesis can never muster more than subjective certainty, and will always be vulnerable to accusations of arbitrariness. Does this matter? It matters because of Hirsch's own concern for validity: his analysis of interpretation is itself a form of meta-interpretation and yet fails the very tests that he is claiming should characterize the practice of interpretation. As a higher level practice, he could argue that its strength lies in its descriptive (rather than normative) qualities but there is an evident contradiction in him putting the case for validity in interpretation when he is unable to validate either the need for validity or the intentionalist route to it that he chooses.

The all too common accusation of arbitrariness levelled against Hirsch is depicted by Meiland as follows:

> We *set up* our own reality rather than find a reality already existing as such independently of us; we do this by deciding to set up the author's meaning as *the* meaning of a work in order to have something to inquire into. (1978: 26)

However, Meiland denies that this proves a lack of objectivity in Hirschian intentionalism, because, 'the author's intention is a reality independent of our will and in that sense is an objective reality about which we can inquire and legitimately hope to reach objective results' (27). He is supportive of Hirsch's quest for cognitivity, but contends that the goal of a single correct interpretation 'does violence to the richness, the multifarious aspects and levels, of many works of art and literature', maintaining that 'physical science is only one possible model of a cognitive discipline' (31). However, Meiland's pluralism-friendly version of cognitive criticism is not as distant from Hirsch's as he imagines. He suggests that we think of an interpretative criterion as 'ruling out wrong answers, leaving all of the survivors to be considered equally correct' and argues that this 'ruling out' should be done on the basis of textual evidence, thereby allowing for pluralism without anarchy (ibid.). But, strictly speaking, Hirsch does not see validation as offering a single, correct interpretation (even if he might occasionally appear to desire this); any answers it gives are hypothetical, always subject to potential revision in the light of new evidence: 'to validate is to show that a conclusion is probably true on the basis of what is known' and 'correctness [. . .] can never be known to be achieved' (1967: 171, 173). Such ascertaining of probability (rather than truth, the goal of verification) will, however, serve to *rule out* certain interpretations even if it will not alight upon an indisputably correct answer. In practice, the differences between them dwindle: Hirsch posits that a text possesses a single correct interpretation but that, even if the critic correctly ascertains this, he or she can never know for certain that they have done so; therefore the 'violence' to the text that Meiland warns of, will never actually come about.

Meiland himself is guilty of some violence in the representation of Hirsch's position. He interprets the statement that 'a word sequence means nothing in particular until somebody either means something by it or understands something from it' (Hirsch 1967: 4) as the much stronger assertion that 'because the text can be made to say anything at all, it does not rule out any particular interpretation' and, moreover, he describes this as 'the basis of Hirsch's entire view' (1978: 33). However, Hirsch never says more than that linguistic conventions and textual evidence cannot definitively determine (*rule in*) an interpretation; he does not say that such criteria cannot *rule out* obviously false interpretations (and, as we have seen, the rule in/rule out distinction is central to Meiland's description of cognitivity). It is a misreading of Hirsch to claim that for him 'the text itself has no meaning' when he states only that it has no *determinate* meaning (ibid.). He does not claim that textual meaning sets no interpretative parameters whatsoever. On the basis of this misrepresentation, Hirsch is vulnerable to a stronger accusation of arbitrariness: namely, that for him there is no intrinsic connection between authorial meaning and textual meaning, because the text has no meaning. In view of this, authorial intention is no better a criterion (or any more 'cognitive') than, say, reader's meaning.

What Hirsch *is* guilty of is the dilution of the ethical force of his argument by not asserting the connection between authorial and textual meaning strongly enough and by offering parallel – and potentially conflicting – arguments for intentionalism, rather than making the ethical and epistemological elements of his argument more obviously co-dependent. But the most fundamental weakness of Hirsch's position is that, if the normative dimension of hermeneutics belongs to the realm of ethical choice, the establishment of universal normative principles (commensurate with universal scientific principles) seems doubtful. Can his theory in fact do more than describe his preferences? Witness his admission that 'interpretive norms are not really derived from theory [. . .] theory codifies *ex post facto* the interpretive norms we already prefer' (1976: 76); this is a stick that his critics have borrowed to beat him with.

Despite Beardsley's criticisms of Hirsch's 'Identity Thesis', that is, the identification of textual meaning and authorial meaning, the tag is misleading in its suggestion of a logical connection between the two (1968: 169). In fact, Hirsch merely argues the desirability of the connection on ethical grounds, not that they are 'one and the same thing' as Beardsley claims, because it suits his view of what interpretation should do (ibid.); he also, with limited success, proclaims the anti-ontological status of his argument. By contrast, P.D. Juhl, with his self-professed 'radical intentionalism', aims at a descriptive analysis of the concept of literary meaning and the ontology of literary texts and concludes that 'There is a *logical connection* between statements about the meaning of a literary work and statements about the author's intention such that a statement about the meaning of a work *is* a statement about the author's intention' (1980: 12).

In claiming that a critic is always already referring to intention when interpreting a text, Juhl builds upon but substantially departs from Hirschian intentionalism. The departure is his assertion that an intentionalist stance is a logical consequence of the nature of texts and, therefore, if a critic appeals to textual features, rules of language, or coherence under a particular interpretation, he is implicitly appealing to the author's likely intention. Intention, for Juhl, is compelling, not because it promises a determinate object of interpretation or is a route to a more 'cognitive' criticism, but because it corresponds to his notion of what it is for a literary work to mean something. In this way he sees himself as evading the accusations of arbitrariness levelled at Hirsch by countering the recommendations proffered by his predecessor with what he claims to be an analytic statement: the logical identity of authorial meaning (intention) and textual meaning. Indeed, he contends that public consensus is as compelling a normative principle as intention, but it is the latter, not the former, which has a logical connection to literary meaning.

Interestingly, Juhl fails to construct any kind of working methodology upon the basis of his fundamental logical claims. He resists the urge to make his descriptive account normative or prescriptive. While holding that a text has only one correct interpretation which is, in principle, determinable, he offers no advice as to how this correct interpretation might, in practice, be determined. His point is rather that every critic who talks about literary meaning is already an intentionalist, on the grounds that an appeal to the text *just is* an appeal to authorial intention and that textual instances of allusion and irony and so on have a necessarily intentional aspect; irony is connected to *use*, not a latent feature of the words themselves. Similarly, an appeal to context is an appeal to the intention of the author who first determined that context, furthermore, 'in excluding an interpretation ostensibly on the basis of the rules of the language, we are implicitly appealing to the author's likely intention, given his knowledge of the rules of the language' (106).

In short, all textual features are the (logical) consequence of authorial intention, because they are intentionally selected and arranged. This is problematic in several ways. First, Juhl fails to account for textual features, effects and meanings not intended by the author, when even Hirsch does not deny that such features exist. Second, his logical intentionalism does not allow for unconscious meanings – which Hirsch tries to explain with his concept of a 'willed type' (a notion which will receive further discussion later in this chapter). Third, and relatedly, Juhl's central thesis rests on certain questionable assumptions about the rationality and competence of the author, and the author's desire to communicate his meaning clearly – he may wish to pervert and obfuscate meaning, in which case his meaning just is this obfuscation. Would the reader recognize this? Perhaps not. This is the final objection levelled at Juhl: that he fails to explain the misreadings which inevitably occur in criticism. On his account, all and any meanings garnered from the text are authorial meanings and are, therefore, sanctioned. In this way, his intended homage to Hirsch has, in its failure to be prescriptive, become an endorsement of unbridled relativism.

Intention and convention

Siebers maintains that 'literary criticism, as the New Critics understood, needs a theory of language, but no viable theory of language can exist in the absence of a human ethics' (1988: 68). Hirsch has his ethics, even if he is sometimes reticent in proclaiming them. What, then, is his theory of language? Rather evasively, he avers that 'sometimes the use of language is uniquely constitutive of meaning; sometimes, apparently, a particular choice of words merely imposes limitations and is not uniquely required for the meaning that is actually willed' (1967: 28). But fundamentally, Hirsch's is a theory of language in the service of the human and he resists a more 'organicist' conception of literary language as leading a life of its own, unhindered by the referential responsibilities of 'ordinary' language; instead, the emphasis is on the human will that selects meaning. Nevertheless, Hirsch must describe the agreed structures *within which* such selection takes place and this stimulates a certain amount of tension between intending will and conventional constraints: what Hirsch calls 'the double-sidedness of speech', the fact that it is both somebody's meaning and yet interpretable by a wider public, and according to public norms (69). This tension is evident in his initial definition of verbal meaning in *Validity* as 'whatever someone has willed to convey by a particular sequence of linguistic signs and which can be conveyed (shared) by means of those linguistic signs' (31).

The potential conflict, then, is between the intending will and the possibilities afforded by the common language (which is not a 'conflict' as such, but which requires, ultimately, that one be privileged over the other); Lentricchia, however, goes further in asserting a more fundamental tension – between 'genesis' and 'structure' – actually *within* Hirsch's conception of intention, suggesting that 'As genesis, intention guarantees original meaning; as structure, it guarantees that all meanings of a given text are governed by an internal legality of rules' (1983: 280). Thus intention is revealed, first, as serving two express purposes within Hirsch's critical system and, second, as possessing a certain conceptual malleability in its potential for location both inside and outside the text, as cause and organizing principle, respectively. The theory of type meanings which constitutes his theory of language works to break down oppositions such as that of genesis and structure, however (Hirsch is more of a deconstructionist than he might like to admit), and it is the type which will be seen to unite 'the particularity of meaning with the sociality of interpretation', in this way encouraging a kind of dialogue between intention and convention and identifying the literary text as, in Hirsch's estimation, both personal expression and public document (1967: 71).

The type theory also guarantees Hirsch's goal of genuine knowledge in interpretation by providing the critic with a stable, determinable object of interpretation. Thus: 'In order that a meaning be determinate for another it must be a type. For this reason, verbal meanings, that is shared meanings, are always types and can never relinquish their type character' (50). It is notable that he appears to argue for a typological theory of meanings, not because of something in the nature of language, but because it is a means to his desired end: validity requires an object (literary meaning) which is both determinate and 'sharable'. There is a danger of circularity here: verbal meanings are 'sharable' and, unlike things, are 'capable of being fully known' because they are types, yet he argues for the typological theory on the grounds that it must obtain if meaning is to be sharable, and so that one can 'subsume or represent two different entities by the same word' (273, 265). He doesn't argue (although he could) that because it must be in the nature of a functioning language system to be able to subsume or represent two different entities by the same word, and because our language system indubitably functions, therefore the typological structure must obtain. Lentricchia reads Hirsch as claiming that 'since all texts are instances of types, all texts are translatable, because types, to be types, must subsume many instances' (1983: 273). This seems like a fair characterization, but is actually flawed in the direction of reasoning that it assumes – rather than arguing that texts just are types, Hirsch argues from the perspective that they must be translatable (if worthwhile criticism is to proceed), and a type theory of texts ensures their translatability. As ever, the avowed emphasis with Hirsch is upon ethical choice rather than ontological necessity.

Hirsch's intentionalism emerges in his statement that '*a verbal meaning is a willed type*'; in this way intention is kept central to the *process* of meaning, a process of co-determination by intending will and facilitating convention (1967: 51). For Hirsch, whose theory must account for both objectivity and ambiguity in the interpretation of texts, a type theory allows verbal meaning to 'be (as it is) a determinate object of consciousness and yet transcend (as it does) the actual contents of consciousness' (49). The appeal of this is that it confirms our pre-theoretical intuitions about literary meaning: first, that a text means something that is in principle determinable (otherwise the problem arises of how to explain the appeal of reading and the point of criticism); second, that it is likely to mean what the author intended; third, that in practice it means more than the original blueprint that the author held for it, that is, it amounts to more than the 'contents' of his consciousness at the time of creation.

The determinacy here arises from a work being 'a particular utterance', the norms governing which are 'definitive and determinate', rather than 'elastic and variable' as the general norms of language are (69). The particular 'norm' governing a literary text is its 'intrinsic genre', which Hirsch describes as 'that type which embraces the whole meaning of an utterance', that is the particular, definitive actualization (out) of the (various) meaning possibilities sponsored by the piece of language in question (71). Based on his assumption (his initial, divinatory guess) of the type of meaning being proffered, the interpreter looks to the traits that he would expect to constitute such a type and so divines the detailed meanings of the text. Thus: 'An interpreter's preliminary generic conception of a text is constitutive of everything that he subsequently understands, and that [. . .] remains the case unless and until that generic conception is altered' (74).

Hirsch does not see the apparent self-confirmability of interpretations as problematic; indeed he thinks that all meaning and understanding is 'genre-bound' (78). However, circularity is avoided because of the reciprocal (mutually constitutive) nature of whole and part which makes understanding (and interpretation) a kind of dialectical process. He notes, therefore, the 'imprecise and variable' nature of the concept of genre:

> A generic conception is apparently not something stable, but something that varies in the process of understanding. At first it is vague and empty; later, as understanding proceeds, the genre becomes more explicit, and its range of expectations becomes much narrower. (77)

This means that the initial assumption of genre is always open to revision and the whole (genre/type) is *constituted by* as well as constituting its parts (traits). In this way Hirsch appears to evade the perils of the hermeneutic circle. Nevertheless, there remains the danger of tautological reasoning if the intrinsic genre is equated with the meaning of the utterance as a whole and not developed any further: to understand the meaning of the utterance you must first grasp the intrinsic genre (which *is* the meaning of the utterance). So, Hirsch maintains that 'there are fewer intrinsic genres than there are particular meanings' and refines his definition of 'intrinsic genre' as follows: '*It is that sense of the whole by means of which an interpreter can correctly understand any part in its determinacy*', and so is not identical with the utterance's particular meaning which 'arises when the generic expectations have been fulfilled in a particular way by a particular sequence of words' (82, 86). Prior to the fulfilment of the generic expectations the interpreter may revise his initial generic guess (the whole meaning) without revising his opinion of the traits (sub-meanings) in evidence, because a particular trait may belong to several different genres, may sponsor several different overall meanings; a change in the whole does not necessitate a wholesale change of parts.

It is not only the interpreter who relies upon the concept of intrinsic genre; it also governs the speaker/author's anticipation of (intention for) his meaning, allowing him to choose and order his words. He first wills the *kind* of thing he is going to say or write, then chooses words which will ultimately express a *particular* meaning:

> His meaning in all its particularity depends on the particular choice of words by which he realizes that type of meaning. [. . .] the further determination of his meaning depends entirely upon his subsequent choice of words and patterns falling within the tolerance of the intrinsic genre. (86)

But what, and how broad, is the 'tolerance' of the intrinsic genre? Which implications are warranted and which are precluded by the generic conception of the text's meaning? Hirsch writes that intrinsic genre is 'the principle by which we can discover whether an implication belongs to a meaning', but he must then devise a further principle by which we can discover whether a particular intrinsic genre warrants a particular implication, since such 'warranting' does not amount to logical entailment (89–90); it is weaker than this, *despite* Hirsch's declaration that 'from the premised type of meaning, the implication follows with necessity' (91). It seems to me that his adoption of the formula 'if the meaning is of *this* type, then it carries *this* implication', while it captures the hypothetical nature of interpretative statements (if . . . then . . .), insufficiently explains the type-implication relationship (ibid.). Can we really so easily deduce one from the other? He is suggesting that we always already know which types foster/support which implications and which implications comprise which types, that all are part of a learned system. Such circularity, despite having a kind of intuitive appeal in the way it characterizes the process of meaning construal, yet fails to provide Hirsch with the validity that he seeks, by giving him some ground or origin outside the circle.

This is the crux of the objection to a type theory of meanings: the divination of intrinsic genre is only possible by a process of comparison with past meanings – the reader or critic recognizes the meaning as being a particular type (and, by extension, knows all its implications) because he has previous experience of this type. The difficulty lies in explaining how anything is ever expressed for the first time, whether there is such a thing as a 'new' meaning (or genre) that is anything more than a combination of the elements of past meanings (what of neologisms, for instance?). Hirsch doesn't offer an empirical account of types, where familiarity with a type requires the subject to have had actual experience of the type thing (or a depiction of it) and he does posit 'provisional, broad heuristic concepts' by means of which new genres come into existence (104). Nevertheless it is still old types which are 'the foundation for new ones', ensuring the communicability of the new types of meaning, which otherwise would be 'radically ambiguous' (ibid.). The coming into being of new type concepts requires 'pre-existing type conceptions' which can be amalgamated or extended (ibid.); there is never a wholly new type. In this way Hirsch stresses the interconnectedness of all types; there is some essential overlapping where traits belong to more than one type and there is no such thing as 'an altogether explicit type' because such a thing 'could be subsumed by only one instance' and would more properly be called 'an individual' (270). Furthermore, types are recognized as much by the traits that they fail to exhibit as by those that they do: a negative, differential account of type meanings that Hirsch develops from his reading of Saussurean linguistics.

The typological structure leads Lentricchia to characterize Hirsch's theory of language as a rejection of particularity in favour of more generalized (generic) meanings, and validity as a restrictive impulse, limiting the interpreter to 'the recognition and articulation of the typological features of a text and the typological self which brought that text into being', such that he 'run[s] up against the limits of cognition' (1983: 268). Does such a restriction amount to a reduction of the text to the merely conventional (and the author to a cipher), to meaning as merely a system of conventions? A more generous interpretation would see Hirsch as aiming at an understanding of particularity (particular meanings which are not immediately accessible) through generic meanings (type concepts which are). It is intention which lends a statement/utterance its particularity, based on the uniquely subjective perspective of the speaker, but despite Hirsch's emphasis on the selective power of the will, such a perspective must always be expressed within the conventions of language if it is to be 'sharable'.

In attempting to effect this reconciliation between individual intention and shared convention, Hirsch is responding to the most damning and efficacious attacks upon intentionalism which foreground the public nature of the text and the apparent inaccessibility of authorial intention: this intention becomes 'sharable' (accessible) if understood as instantiated through some already known type. According to Lentricchia, the type operates at the level of (primary) intention for Hirsch, so meaning is always already convention-bound: 'verbal intention itself, fully conscious, fully possessed, remains the prelinguistic origin or place of residence in the subject of a type meaning' and the type is sufficiently restrictive that the meaning of the finished text fails to exceed (escape the boundaries of) this 'basic typological form' (1983: 272, 273). This subsumption of the particular by the general threatens to contradict Hirsch's earlier assertions of the originality and singularity of literary meaning. To detract from the apparently dictatorial power of linguistic conventions, Hirsch describes them as being of a 'broadly social character' and therefore adaptable and speaks of 'proprieties' rather than 'rules' (1967: 93): the speaker/author may choose the extent to which he follows such conventions, but how much choice do authors really have if they wish their meaning to be understood? Hirsch's get-out clause is in the chronology of the creative (and the interpretative) process, for 'it is the speaker who wills the particular intrinsic genre and, *having done so,* is constrained by its proprieties' (94; my emphasis).

In this way – as the originator of meaning – the author retains a relative autonomy from linguistic conventions; he is not quite their master, but does at least choose which conventions he wishes to be mastered by. Intending will 'lies at the heart of what a genre is', writes Hirsch, 'the unifying and controlling idea in any type of utterance, any genre, is the idea of purpose', and this unifying purpose is the author's 'notion of the type of meaning to be communicated', without which animating 'idea' the genre is 'mute inert matter' (99, 101). Nevertheless, the author's meaning is never realized or understood as unique or 'pre-typological', even by him – indeed this is *why* Hirsch holds authorial meaning to be sharable, because it is conceived and constituted through the type meanings with which we are all familiar, because it is never 'particular' in the absolute sense of that word. Otherwise, the originary power of selection and the theoretical adaptability of linguistic conventions together would constitute a difference that failed to make a difference for interpretation *if* we were unable to reconstruct or recognize the particular animating idea (intention) of the author and so discover which implications deserved emphasis and which did not.[2]

In Hirsch's subsequent analyses of the issues surrounding literary intention, we can see the development of the ideas first expressed in *Validity in Interpretation.* Meaning as an intentional 'object' is no longer 'fixed and immutable' because such objects are 'never completely present to the mind' (1984: 202, 203). A meaning may have implications or aspects which were not intended or realized by its author and such implications are more difficult to determine than, say, the hidden sides of physical objects, which may be deduced by means of standard physical principles. No such determinate and determining principles exist in the realm of meaning. While the text's 'originating moment in time' fixes 'the *principles* of further extrapolation', this extrapolation is not a straightforward business and the principles themselves are indeterminate (204). The clarity of his earlier distinction between meaning and significance suffers as a result; whereas, in the 1960s, Hirsch saw meaning as deriving a reassuring fixity from historical (i.e., authorial) intent (and it was this which marked it out from 'mere' significance), now he replaces this stable, determinate, historical intent with the more nebulous idea of a 'variable, future-oriented intent' (205). In order to figure this 'future-oriented intention' as *both* self-identical and variable (a difficult task for Hirsch) he describes it as 'an explicit plan with areas of inexplicitness' and the inexplicitness is seen to arise from the nature of meaning and understanding as diachronic process (as opposed to *fait accompli*) (206). Thus we may experience future-oriented intentions even in face-to-face conversation, because the listener receives and hears the words at a point (marginally) later than the speaker utters them. In the case of writing there may occur 'an indefinitely long pause between the speaker's present and the listener's present' (ibid.). It is this pause which Derrida has characterized as a kind of aporetic space, leading to the necessary possibility of infelicity and misunderstanding in communication; this will be the subject of a later chapter. Hirsch, however, clings to his notion of validity, despite the possibility of miscommunication, by suggesting that it is in principle possible to determine which are 'genuine future fulfilments' of the original intention, and which are not (207).

This does not appear, then, to be much of a departure from his earlier discussion of meaning–implications, but here he uses the analogy borrowed from logic of the extension of a concept to cover all past, present and future instances of a thing; in this model, a meaning is a genuine 'meaning–fulfilment' if it is subsumed by the original concept(s). This suggests a newly radicalized notion of conceptual fluidity and evolution on Hirsch's part, while also referring back to the generality of his type theory of meaning: it is likely to hold if the original meaning–intentions are *broadly* conceptual, that is, if they are general concepts likely to exist into the future. The paradoxical nature of a concept as 'both an "internal" generality and an "external" array of things embraced by the generality [. . .] both an "intension" and an "extension"' facilitates Hirsch's use of it as a metaphor for the meaning process (210). He also co-opts Kripke's work on reference [in *Naming and Necessity* (1980)] into his argument, claiming that acceptable meanings are those which fall under the intended reference, all of the implications of which the author/speaker may not be able to predict. This focus on the referentiality of mental states rather than mental content allows for and explicates meaning which changes *within the boundaries* of the possible reference (of the original intention). The necessary ambiguity of reference gives good scope for changing meanings and confirms our intuitions regarding authorial fallibility or inexplicitness – 'we never certainly refer' (1984: 221) – while also placing an emphasis on externality which serves to ward off objections of a psychologistic nature.

As previously noted, the revised conception of meaning does create problems for the meaning–significance distinction. That which, in Hirsch's earlier account, fell under the title of significance – future implications and applications of meaning – is now seemingly promoted to the status of meaning proper, if it falls within the boundaries of the possible reference; if it falls outside this, then it appears to be disqualified outright. Nevertheless, Hirsch labours to preserve the distinction, maintaining that 'clarity will continue to be served by distinguishing between what stays the same and what changes in different interpretations' (ibid.). The practical problem of how to differentiate between meaning and significance remains, however, given the new assertion of the indeterminacy of the extrapolatory principles. While Hirsch has drawn closer to Gadamer (the target of his criticism in *Validity*) in his embracing of application as a valid part of meaning, he still wishes to distinguish his position from Gadamer's. The distinction, though, is primarily in motivation, and Hirsch will not surrender his ethical objectivism: while Gadamer argues for 'the necessity of differentness of meaning', Hirsch rejects this – on ethical grounds – and argues instead for 'the possibility of sameness of meaning, in different applications of the text' (1984: 214). He opposes his 'historicality' (the principle that meaning *may* remain the same over time, if regarded as an 'historically determined object') to Gadamer's 'historicity' (the idea that meaning *must* change over time) (1984: 216). The advance on *Validity* and *Aims* is the weakening of his original declaration of the ineluctable sameness of meaning, which is now the positing of its mere 'possibility'. In doing this he opens the door, cautiously and with trepidation, to the possibility of Gadamerian-style 'differentness' of meaning, meaning which has not been *explicitly* – this is the crucial qualification – intended by the author. Furthermore, he still ostensibly resists giving an ontological argument, and suggests that we can *choose* to treat meaning as Gadamerian or Hirschian, that we can choose to see the influence of history negatively or positively. The ethical choice, for Hirsch, is to see the text as an 'historically determined object', but nothing beyond his particular ethics – as evinced by his concern for 'the value and credibility of humanistic scholarship' (218) – compels us to do this. And now that 'new exemplifications' and even 'minor conceptual adjustments' can be part of what Hirsch still terms 'self-identical meaning', the Hirschian critic could be forgiven for experiencing a certain conceptual confusion (224).

The author and authority

Whatever the tentative radicalism of Hirsch's position – his attempt to provide an account of literary meaning which accounts for all its constituent parts (whether private and mental or public and conventional) and the paradoxical nature of their arrangement, and his later shift in emphasis away from a focus on creative origins – he fails to fully establish and explore a conception of meaning which is not tied to the author, that is to say, to some single, identifiable, determinable creative consciousness. Foucault writes that:

> Criticism has been concerned for some time now with aspects of a text not fully dependent on the notion of an individual creator; studies of genre or the analysis of recurring textual motifs and their variations from a norm other than the author. (1977a: 126)

Yet interestingly what Hirsch tries to offer is a study of genre (i.e., a basically formalist approach to the text) that is yet dependent upon this notion of an 'individual creator', which sees genre as willed, as intentional. Intentionalism for Hirsch will always be authorial intentionalism. The ethical timbre of his work rests upon this reverence for authorship, but he fails to recognize the ramifications for the discipline of criticism: which ramifications, to my mind, include the limitations of an authorialist position, as far as a developed conception of intention is concerned. Validity requires the stable reference of authorship; when he does suggest a weakening of the link between author and meaning, the validity thesis collapses, because it has been built upon the foundation provided by this link; it could, however, have been differently constructed. In fact, as we have seen, he weakens the link only incrementally: the creative consciousness still decides the '*principles* of further extrapolation' of meaning, if not the meaning itself (1984: 204). Thus the author keeps his authority (to set boundaries or determine the 'horizon' of possible meanings), if not the absolute exclusivity of his claim (to determine meaning absolutely and for ever).

Authorial intentionalism is presented as a commonsensical approach to criticism by Hirsch: on the first page of *Validity* he reveals his admiration for 'the sensible belief that a text means what its author meant' and offers a self-consciously heroic 'defence' of the author, who has suffered an ignominious 'banishment' in literary criticism since the 1960s (1967: 1). Although his 'meaning is an affair of consciousness' declaration need not tie meaning to the author's consciousness in particular, he insists that it is morally more justifiable to adhere to the author's meaning than to the critic's. There is no logical reason for this – either could provide a stable normative criterion for interpretation, and authorial meaning is not necessarily the 'best' (most aesthetically pleasing, most cohesive, most productive) meaning – but it fits with his emphasis on the singular meaning of a text to favour single author over multiple critics (with their multiple readings) for, as we have seen, pluralism spells the end of cognitive criticism as figured by Hirsch.

Defending the author against the anti-intentionalist charge that meaning changes over time, Hirsch calls upon his meaning/significance distinction, insisting that it is only the latter which changes: 'Significance always implies a relationship, and one constant, unchanging pole of that relationship is what the text means' (8). To the further objection that 'the meaning of a text changes – even for the author', Hirsch responds that an author changing his mind is not sufficient to change the original meaning of the words, which remains as it was, regardless of the author's new opinion of it (6ff.). However, he fails to differentiate between the two hypothetical situations that he posits: first, the author changing his mind about what he *now wants* his words to mean – 'from now on they shall mean something different' – that is, different from their original meaning, X, but that X remains unaffected, even though his feeling about X has altered; second, the author changing his mind about what he *meant at the time* – 'I thought I meant X, actually my words meant Y' – Hirsch fails to explain the effect upon X if X was always a miscommunication and misconstrual. This second situation poses a problem for Hirsch. He is right that the author 'could report a change in his understanding only if he were able to compare his earlier construction of his meaning with his later construction' (9); but the question concerns the status of the 'earlier construction' – whether it ever constituted meaning (rather than significance), if the author says he *never* meant what he was construed (by himself and others) as meaning.

Another way in which Hirsch attempts to explain away authorial confusion or ignorance – responding to the accusation that the author doesn't know what he meant, that the critic knows better – is in the introduction of another distinction: that between subject matter and meaning. The author may be insufficiently acquainted with his subject matter while a reader or critic of a subsequent age may be more conversant with or better informed about a subject, but this does not amount to the author being ignorant of what he means *by* what he says; the truth or falsity of what he says is irrelevant. However, Hirsch advocates the reconstruction of authorial meaning in order to provide interpretation with a determinate object (something which he believes the text itself cannot do because a particular word sequence can stimulate a number of different meanings), and much of the earlier discussion suggests that authorial intentionalism may actually fail to provide this, that meaning is always in a kind of flux and may not be assured, even for the author.

In response to the accusation that the author's intention is, in any case, inaccessible, Hirsch asserts that most authors do not set out to be 'obscurely autobiographical' and all must work within publicly known genre conventions, adding further that 'the irreproducibility of meaning experiences is not the same as the irreproducibility of meaning [. . .]. Meaning experiences *are* private, but they are not meanings' (16). Meaning is sharable, accessible – thanks to its typological nature – and knowledge of meaning need not incorporate knowledge of the entire contents of the author's mind at the time of creation, that is, knowledge of his 'meaning experience'. The question remains why Hirsch should think that the only worthwhile and, more importantly, the only truly determinate meaning carried by a text is that given to it by its author. While one may think that a work must have an author to be meaningful (must be an artefact), this does not entail the conclusion that its meaning must be all and only the author's meaning; this conclusion is Hirsch's premise. Thus Beardsley alleges that, rather than admitting 'textual meaning can go beyond authorial meaning', Hirsch 'tries to stretch the concept of *will* far enough so that whatever the text *does* mean can be said to be "willed" by the author' (1970: 21). Hence Hirsch's sanctioning of 'unconscious' meanings which may have been 'present in another region of [the author's] mind' or present in his 'subconscious mind' (1967: 52) but are yet 'willed'; 'unconscious implications [. . .] belong to the intention taken as a whole' and so constitute a valid object of interest for the critic (1967: 221). However, the distinction that this proposes between unconscious (willed) meanings and symptomatic (involuntary, unwilled) meanings is difficult to sustain, and there is an intuitive appeal to Beardsley's rejoinder that 'whatever is unwitting is unwilled' (1970: 21).

According to Meiland, Hirsch's 'further step' (from his supposition concerning the artefactual nature of texts) is to view the text as a kind of *communication* or message; such a conception of meaning, says Meiland, entails that 'a work cannot have a meaning by itself independently of the author' (1978: 39). This is a false inference on Meiland's part, however. The meaning-as-communication view may dispute the *relevance* of apparent textual meaning where there is evidence that the author/speaker meant something quite different, but generally textual meaning will be seen as a vital *clue* to author/speaker meaning, on the assumption that the author is working within certain linguistic conventions. A meaning-as-communication theorist allows that a work has meanings, knowledge of which can be achieved independently of knowledge of its author's intention and meanings which may in fact be 'unauthorized', but such a theorist is likely to be interested in apparently unauthorized meanings *only insofar* as they lead us to an understanding of what the author *was* trying to communicate. To see sentences as purposive is not necessarily to see them as meaningless when stripped of purpose – when the guiding intention cannot be known.[3] Nevertheless, it may be asked why the privileging of the author retains its appeal for Hirsch in situations where the direction of interpretative inference is from text to author, not vice versa. Furthermore, in order to retain the author at the centre of the interpretative enterprise, Hirsch has had to undermine that author's supposed authority, thus mimicking the move of anti-intentionalists who aim to discredit the author and confer authority upon the text itself. Hirsch agrees that the author may *not* be the most authoritative reader of his own work, that he may change his mind about his own meaning–intentions, that he may not be aware of all the implications of his words, and yet insists that we continue to subscribe to the ethical ideal of our responsibility to authorial meaning, which responsibility comes to seem more and more arbitrary.

The evolution of his thinking on this subject is evident in his revision of the definition of meaning in *Aims*, which clings to the ideal of semantic determinacy while tentatively renouncing the author as the guarantor of this determinacy. He now writes: 'meaning is the determinate representation of a text for an interpreter', claiming that 'the enlarged definition now comprises constructions where the authorial will is partly or totally disregarded' (1976: 79–80). Hirsch further undermines the author in *Aims* – albeit unwittingly – by reiterating the uncertainty of the interpretative process when it is authorial meaning that is its object. Claiming that it 'cannot be determined with absolute certainty' whether the meanings 'actualized' by the reader 'are also meanings intended by an author', he yet says that 'the reader is in fact free to choose whether or not he will *try* to make his actualized meanings congruent with the author's intended ones' (8). But a reader cannot even attempt such a congruence if he cannot first know with certainty what the author means, that is, cannot first know *with what* he is attempting a congruence. He can only know with certainty what the author doesn't mean and can ensure only that he doesn't actualize *obviously false* meanings. The new, more capacious role for the interpreter – all but excluded from *Validity* – and Hirsch's unwitting questioning of the accessibility of authorial intention leaves the author as just another player in the interpretative process. Does Hirsch no longer hold 'the sensible belief that a text means what its author meant' (1967: 1)?

William Irwin claims that Hirschian intentionalism's lack of 'a theory of authorship' is its 'major deficiency' and he aims to supplant Hirsch in the (somewhat diminished) ranks of intentionalists with his own theory of 'Urinterpretation', central to which is the notion of the 'Urauthor' (1999: 39, 11ff.). 'Urinterpretation' is defined as an interpretation 'that seeks to capture the intention of the author, though not necessarily his understanding' and the German prefix 'ur' connotes origins and also refers to the word 'Urheber' – author (11). Like Hirsch, Irwin sees a text as a kind of communication, explaining that 'inasmuch as the author's text is an intentional effort to communicate [. . .] his intention is indeed worth recapturing', although his understanding of his text may not be, as this may be flawed (12). This last point seems hardly worth making: the unquestionable authority of the author's understanding of his own work is not something which even the most ardent intentionalist attempts to maintain, despite anti-intentionalist comments to the contrary.[4] Like Hirsch, Irwin also views interpretation that is not intentionalist in its focus (what he calls 'non-urinterpretation') as pointing to 'significance' rather than meaning; such interpretation is also 'potentially unethical' (ibid.). So far, it is only in his terminology that Irwin has succeeded in distinguishing himself from his more illustrious predecessor.

However, it is in his conception of authorship that Irwin hopes to ground his intentionalism, and this is worth examining. It is a conception that exists primarily in relation to the discussion of authorship by Barthes, Foucault and Nehamas. For Foucault, the coming into being of the 'author' represents 'a privileged moment of individualization in the history of ideas, knowledge, and literature, or in the history of philosophy and science', which is open to challenge now that 'the writing of our day has freed itself from the necessity of "expression"', a 'reversal' which 'transforms writing into an interplay of signs, regulated less by the content it signifies than by the very nature of the signifier' (1977a: 115, 116). Such writing is no longer tied to 'the exalted emotions related to the act of composition or the insertion of a subject into language'; rather it is 'a game that inevitably moves beyond its own rules' (116). In other words, there has been a period of history (pre-Enlightenment) when a text was not tied to a particular author, when it was not a product that could be owned by a particular author, and such a period *could* come about again; unlike Barthes, Foucault does not claim that such a 'death' of authorship has already occurred, although he does appear to concur with the abandonment of the humanist conception of the subject as the locus of all meaning (and of all meaning as possessed of a human source). What Foucault is noting is the disappearance of the peculiarly modern conception of the author: a lesser sort of banishment. Nevertheless, 'An author's name is not simply an element of speech [. . .]. Its presence is functional in that it serves as a means of classification. A name can group together a number of texts and thus differentiate them from others (123). It is this functional role of authorship that we are dependent upon, says Foucault, and which will be difficult to dispense with: namely, the author as 'author-function' rather than historical person, as a classificatory or limiting principle in interpretation (Foucault is interested in the 'repressive' effect of such authority) and as an indicator of the special status of certain kinds of texts (127). The 'author-function' is a construction, made up of aspects which are 'projections, in terms always more or less psychological, of our way of handling texts' and the construction is historically relative, it is not 'defined by the spontaneous attribution of a text to its creator, but through a series of precise and complex procedures' (127, 130).

Irwin's objection to Foucault's conception of the author-function concerns its characterization as repressive, and he argues that 'it is not the mere existence of the mental construct, known as the author function, that is repressive' but rather the 'author function constructed in strict accord with the historical author-as-person' (1999: 21). But this identification of author-function and historical person is something that Foucault explicitly avoids, asserting that the author-function 'does not refer, purely and simply, to an actual individual insofar as it simultaneously gives rise to a variety of egos and to a series of subjective positions that individuals of every class may come to occupy' (1977a: 130–31). Nevertheless, both Irwin and Alexander Nehamas aim to avert Foucault's allegations of the repressive nature of authority by refining the distinction between historical person and author-function. The 'Urauthor' is Irwin's intentionalist author-function and the hypothetical nature of this constructed author figure is intended to represent the fact that 'all we ever have is a *more or less accurate version* of [the historical author] as related to his text, an author construct' (1999: 29; my emphasis). This is not as much of an advance on Hirschian intentionalism as Irwin would like to think. In his most recent writings on the subject of authorship, Hirsch himself subscribes to the Foucauldian belief in the necessity of the author-function, but puts his own intentionalist spin on it as 'an indefeasible *empirical* feature of language which ensures that intention will not go away', adding further that 'there is no empirical requirement that the provider of the author function – the begetter of the meaning-determining intention – has to be the original author(s)' (1994: 551).

This signals a new radicalism in Hirsch's work which sees him neither 'submitting' to, nor 'repudiating' original intent and so embracing a theory of authorship which allows for the presence of both original intentions, which 'are not [. . .] limited to original meanings', and future applications, that is, an author-construct, which is itself subject to change and reinterpretation and which is, like Irwin's 'urauthor', a *more or less accurate version* of the historical author (555). In fact, an earlier, more tentative version of the author-function is in evidence in the 'speaking subject' attributed to every text in *Validity*.

This 'speaking subject' is notably not 'identical with the subjectivity of the author as an actual historical person' but only with that 'part' of the author that determines verbal meaning – hence, Hirsch's stipulation that we do not need to reconstruct everything in the author's mind at the time of writing; such total reconstruction would constitute an attempt to reconstruct the historical person (1967: 242).

Nehamas' particular distinction is between *writer* (historical person, and the text's efficient cause) and *author* (the producer of the text as we construe it, the text's formal cause), thereby suggesting an ontological distinction between the two (1987). This version of the author-function is what Nehamas elsewhere calls 'the postulated author' (1981) and is 'an agent hypothesized to explain the features of a text', occupying the same historical context as the writer, yet distinct from him (Stecker 1987: 263). The postulated author is an enabling rather than a repressive figure, and is useful to the extent that it explains the hypothetical nature of interpretations – something that Hirsch, as we have seen, is keen to maintain, despite his objectivist manifesto. Like Hirsch's notion of intrinsic genre, based on an initial divinatory guess, the author hypothesis changes as we interpret, as more information is forthcoming; thus, Nehamas claims, the identity of the postulated author evolves – although Stecker more plausibly reads this as the evolution of our beliefs *about* the author (264). The key difference is that the author-function is characterized as a kind of formal (textual) feature, and is not seen to possess psychological states: Nehamas' radical monism appears to aim at an author-centred criticism that yet denies the presence of intention. However, the writer/author distinction is difficult to maintain, as his construction of the author-function appears to take its possible parameters from the model of the historical writer; the two are therefore not properly distinct. The author-function is described as 'a character the writer could have been' (285). It is not clear why Nehamas describes the reader's hypotheses as being directed at the author-function rather than the writer: we hypothesize about the author-function which is itself an imperfect rendering of the historical writer; why not cut out the intermediary term?

Irwin's 'urauthor' also suffers from an unhealthy dependence upon the historical author, its very prefix suggesting 'that we should go back to the origin in forming this figure' and the best evidence available for its construction being the 'likely intentions' of the historical author (1999: 30). Like Nehamas, Irwin asserts the ontological separability of historical person and construct – the latter is not 'real' but only a 'thought content' (31) – while also suggesting their mutual dependence: 'To recognise an entity as a text is, in part, to recognise it as having an author (a historical producer) and to recognise an author is, as we shall see, to recognise an author construct', and 'to give thought to the author is to form an author construct' (ibid.). In the light of this it is difficult to see how the author-construct amounts to anything more than an admission of the hypothetical nature of our grasp of the real, historical author. The fact is that the 'construction' still '[resembles] the historical producer as closely as possible in all relevant ways', it is simply acknowledged as being provisional and approximate (61). The admission of such inextricability implies an unnecessary proliferation of terms (and entities) here: the author-construct is the author by another name, despite Irwin's nod to the language of the text and its context. If the author-construct is merely *what we can know of the historical author* then the utility of Irwin's urauthor construct is entirely descriptive and the concept lacks the specifically normative power that he claims for it throughout (and which he characterizes as his amelioration of Hirschian intentionalism). In fact, although Irwin berates Hirsch for omitting to detail the role of an author-construct, Hirsch's system benefits from the much more explanatorily useful concept of the intrinsic genre, which more effectively accounts for the piecemeal progress of interpretation, the necessary revision, the interaction of part and whole.

All of the earlier mentioned points serve as a reminder both of the importance of a theory of authorship – or, at the very least, of some theory of *agency* – to any account (intentionalist or otherwise) of literary meaning and of the perceived failure (according to the late-twentieth-century critical orthodoxy) of accounts of intention which are weighed down with too romantic a reverence for the author as a stable, normative principle. Seán Burke maintains, nevertheless, that 'the ancient chimeras of origin and authorship reassert themselves in the very gestures that seek to have done with origin and authorship', and these 'gestures' (in the form of structuralist and post-structuralist philosophy, the anti-intentionalism of the 1960s and 1970s) will be the subject of a later chapter (1995: xvi). What Burke sees as both apotheosis and nemesis of the tradition of impersonality in literature is productive of some keen ethical conundrums, for any ethical system both requires and resists the violence of the semantic determinacy which authorial authority implies; that is to say, a non-judgemental ethics is not an ethics at all. The author is a god to be disposed of, says Barthes, a redundant transcendental signified, Derrida concurs. Such dystopian visions of the author as a dictatorial and repressive figure, the chosen interpretative criterion of a particularly pernicious variety of critical monism, lead Barthes to see author-based criticism as 'the epitome and culmination of capitalist ideology' and he laments 'the image of literature to be found in ordinary culture [which] is *tyrannically* centred on the author, his person, his life, his tastes, his passions' (1977: 143; my emphasis).

Foucault offers a more moderate opinion, advocating that 'the subject should not be entirely abandoned. It should be reconsidered, not to restore the theme of an originating subject, but to seize its functions, its intervention in discourse, and its system of dependencies' (1977a: 137). But even this implies that for a version of intentionalism to be acceptable in the current critical climate, it must sever its ties with the author-as-originator, and consider more radical, less subject-centred or subject-specific, attributions of intention and agency.

Contemporary versions of Hirschian intentionalism

As we have seen, Irwin's 'urinterpretation' fails to effect this severance, instead retaining the author as dictator of meaning, with the text comprising merely 'an important clue' to this meaning, arguing further that 'texts have no meaning in and of themselves' (ostensibly a stronger point than Hirsch's claim that meaning is 'an affair of consciousness'), 'rather they act as clues, indicating an author's meaning' (1999: 39, 42–43). For Irwin, a text without an author is not a text, properly speaking. In addition, he envisages a broad, amorphous category of 'texts', incorporating legal documents, poems and all literary texts, and insists that all should be read 'intentionally' – notably failing to remark upon the greater claim to an intentional reading that a legal text such as a will could have. There is also the suggestion of circularity in this, as Irwin's characterization of something as a text seems to rest upon the intentional nature of its construction. In any case, such a description tells us little about the entity in question, as texts are not alone in being intentional constructions and, indeed, may not even be this – although Irwin's refusal to admit into the category of 'text' such things as computer poems and unattributed sentences implies that he rejects, without argument, the possibility of intentionless meaning.

A similar rejection is at the heart of Steven Knapp and Walter Benn Michaels' post-Hirschian intentionalism. They insist that, in the intentionalism versus anti-intentionalism debate, 'there are no theoretical choices to be made' and 'it doesn't matter who wins', because neither party 'can really escape intention' (1985a: 18). While this may appear to echo Juhl's logical intentionalism (remember his claim that an appeal to the text *just is* an appeal to authorial intention), there are some key differences. For Juhl, words possess meaning even if abstracted from intention; in this way he makes a distinction between linguistic and utterance meanings (between *langue* and *parole*, respectively) and chooses to treat a text as a kind of utterance or speech act. For Knapp and Michaels, there is no question of choice, and they dispute the linguistic meaning/utterance meaning distinction on the grounds that language that is not a speech act is intentionless and, therefore, is not language at all but rather a random, meaningless sequence of marks. Are they suggesting that all language, in order to be language, must be a speech act, that language does not have meaning outside of or prior to its use by somebody or other? Apparently so. Of speech, Knapp and Michaels write that: 'It is not true that sounds in themselves are signifiers; they become signifiers only when they acquire meanings, and when they lose their meanings they stop being signifiers' and 'what reduces the signifier to noise and the speech act to an accident is the absence of intention' (1985a: 23).

Despite Hirsch's privileging of intention as an animating force and his description of textual marks as 'mute signs' when considered in the absence of intention, he never rejects outright the possibility of intentionless meaning.

Regardless of Beardsley's labelling of his thesis as the 'Identity Thesis', and Hirsch's early identification of meaning and authorial intention, he proceeds to distinguish between the two, in asserting that the former is discoverable through the latter. Holding firm to the 'identity thesis', it is this 'method' that Knapp and Michaels criticize, claiming that: 'Once it is seen that the meaning of a text is simply identical to the author's intended meaning, the project of *grounding* meaning in intention becomes incoherent' (1992: 51). One cannot be 'grounded' in the other because they are one and the same. As interpretation progresses, the advent of new information does not represent the 'addition' of intention, as Hirsch posits, because intention is already present, therefore 'the only thing added, in the movement from indeterminacy to determinacy, is information *about* the intention, not the intention itself' (1992: 53). However, if this intention is not determinable, how are Knapp and Michaels even to comprehend what they are reading, bearing in mind that intentionless meaning, for them, is not meaning at all? They will, undoubtedly, read it according to some animating intention (their own) but as they connect intention with the origins of a work – proclaiming that 'all meaning is always the author's meaning' – possible lack of information (the opacity of intention) is a greater problem than they concede (1985a: 14–15). Is intention for Knapp and Michaels reducible to *information about* (authorial) intention?

This brings us to their famous example of the 'wave poem': words 'written' in the sand on a beach, which apparently comprise some lines from a Wordsworth sonnet, but exist in the absence of a discernible author. If it is a process of erosion, and not some hidden human figure, which has created these marks in the sand, then Knapp and Michaels conclude the marks to be (the product of) a 'natural accident' which merely resemble words, they are 'accidental likenesses of language' (1992: 55). What they are not then able to explain is how the reader who stumbles across such marks is able to 'make' sense of them and so recognize their likeness to Wordsworth's words: if he *finds* the marks meaningful, why are they not so? The difference *for him* between the situation in which the marks represent a speech act and the situation in which the marks are the product of a natural accident is negligible; his experience, either way, is the same, but their theory fails to account for this. Iseminger's astute reading of this is to locate intention at the level of the initiation of conventions; he sees the wave poem as showing little more than that 'the conventions that constitute certain types as items in a language are themselves inconceivable except as the product of intending beings' (1992b: 88). This is a weaker requirement for meaningfulness than that advocated by Knapp and Michaels, and it serves to explain how a wanderer on the beach could understand *as meaning* the words in the sand (which are now seen as meaningful according to intentionally instituted conventions, within a shared language), but it is not sufficient for Knapp and Michaels.

A more intuitively appealing claim than the one concerning the impossibility of intentionless meaning is that there can be neither *expression* nor *interpretation* without intention; as far as the former is concerned, the sea surely did not 'express' the marks, because the sea has no intentional/illocutionary force, but the question is rather: can there be (meaningful) language that is not expression? Their very example suggests that there can be – although, confusingly, they class 'the living sea' as an 'agent capable of intentions' which implies that it could be capable of 'expression' (1992: 55). This suggests that their theory of the intentional nature of meaning is acceptable only if we first presume a more diffuse idea of agency, such that winds and seas and other natural forces are capable of an agency akin or comparable to human agency. Unfortunately, this is not an idea that Knapp and Michaels carry very far. The 'nonintentional effects of mechanical processes' that Knapp and Michaels want to exclude from the realm of meaning are yet the products (and even the producers) of a kind of agency; what they lack is consciousness (ibid.). As for interpretation, to be meaningful it does require an animating intention, but it is likely to be the reader's intention as much as the author's that does the work, and they do not allow for this. In fact they criticize Hirsch for his 'peculiar habit' of positing a choice for the interpreter between author's meaning and reader's meaning, arguing instead that 'there is no choice, ethical or operational, to be made' (53). For Knapp and Michaels, meaning is not only a matter of consciousness, it is entirely a matter of the consciousness of the author. Linguistic conventions, then, are relevant only insofar as they 'provide clues to the meaning the author intends', and even the extent of this relevance is queried, because 'the rules of the language do not provide a range of meanings that are necessarily closer to the author's intention than the range of meanings provided by any other set of rules' (62).

However, there is a contingent if not a necessary connection between the rules of the language and the author's meaning (bearing in mind the author's desire to be understood which leads him to operate within the bounds of those rules), which Knapp and Michaels appear to overlook. They reject Hirsch's view of linguistic meaning as providing a (limited and limiting) set of possible meanings from which the author selects his particular meanings, and so imply (erroneously) that words can mean just whatever the author intends them to mean. From an intuitively appealing premise – the inextricability of meaning and intention – they have derived a number of counter-intuitive conclusions: predominantly, the identity of meaning and intention. This is, says Dowling, 'a kind of too-hasty conceptual elision' and ignores the 'logical independence' of each term from the other (1985: 89–90); entailment of one by the other ensures their inextricability but actually precludes their identity. The result is a descriptive account of literary meaning which is truly anti-theoretical (as they desire it to be) in the impossibility of its application and Knapp and

Michaels gleefully but frustratingly conclude that 'the idea of intention is useless as a guide to practice' (1985b: 101).

While Knapp and Michaels offer the 'identity thesis' that Hirsch never did, Gary Iseminger offers a self-consciously 'Hirschian argument' (1992b: 78); unfortunately, it is subject to the same accusations levelled at Hirsch by earlier critics. Following on from the purportedly uncontroversial assumption that where there are two directly contradictory readings of a poem which are both apparently warranted by that poem, only one will be a true reading, Iseminger then introduces authorial intention as the third premise of his argument, but he does so quite arbitrarily: 'If exactly one of two interpretive statements about a poem, each of which is compatible with its text, is true, then the true one is the one that applies to the meaning intended by the author' (77). However, it is not the case that authorial meaning in particular is that which either entails or is entailed by the determinacy of meaning which is Iseminger's goal. He suggests a logical connection between authorial meaning and 'true' interpretation when the most that can accurately be said is that an interpretation based on authorial meaning is desirable, on ethical grounds. This, though, is not an argument that Iseminger explicitly offers – perhaps his Hirschian argument rests upon Hirschian ethics which he feels no need to reiterate. His premises, however, do not allow him to conclude that 'the truth of [interpretive] statements [. . .] is internally connected with facts about the author's intention', as he has demonstrated no such internal connection (80).

The assumption of determinacy is itself questionable, an assumption contained in the second premise that 'exactly one of the two interpretive statements' is true (ibid.). Again, this is taken as given, despite his awareness of arguments against bivalent truth values, including that propounded by Margolis in this same volume of essays, of which Iseminger is the editor. Margolis' 'Robust Relativism' defends the legitimacy of 'nonconverging interpretations', taking *plausibility* rather than *truth* as the key criterion, and arguing that it is the better of the two because of the 'special interests' of the discipline of criticism and the nature of artworks as culturally emergent entities (in the way that they are formed over time, and never completed) (1992: 41, 44). If Iseminger's two statements are both compatible with the poem, then it could be argued (contra Iseminger) that both are, at least, plausible or even that neither is true. In addition, Iseminger himself concedes that the words in the particular order in which they appear in his exemplificatory poem 'can be used to express either one [of two distinct wishes] (which is not to exclude the possibility that they might be used to express *both at once*)' (1992b: 78; my emphasis). Yet, Iseminger relies unquestioningly on an acceptance of the laws of excluded middle (of two truly contradictory statements, at least one must be true) and non-contradiction (at most, one of the two contradictories can be true), thereby assuming that literary texts are subject to such 'evidently fundamental logical laws' and that literary critical statements are of the kind that can be (absolutely) 'true' or 'false' in the way that they are required to be if such laws are to apply (83). Margolis contends that this might not be the case; I would suggest that it is non-converging rather than antithetical interpretations which trouble literary criticism, and in these cases it is not logical laws which will assist us (non-converging interpretations may not be directly contradictory) but a more general criterion of plausibility, such as that proposed by Margolis. In literary criticism it is relevance rather than rightness that rules the day, and this accounts for the discrediting of such Hirschian notions as 'validity' and 'cognitive' criticism.

Unlike Hirsch, Iseminger does base his intentionalism upon such ontological considerations as the differences between the 'work' and the 'text', proclaiming that 'the text is not the poem' (1992b: 81). It is because the literary work is 'not identical to a text' that, presumably, the former should be read as intentional, an artefact, and so as determinate, while the text may permit a range of possible meanings (i.e., all that are warranted by the relevant linguistic conventions) (ibid.). Iseminger urges us to 'think of the poem as in part *constituted* by something like *willed* or *intended* meanings from among those compatible with the text in order to defend such a principle as premise 3' (84). But why must one think of the poem in this way? If it is only to defend premise 3, then Iseminger is guilty of an arbitrariness from which his claim that premise 3 is 'an ontological principle rather than an epistemological one' hardly absolves him (85). To defend premise 3, one must see the poem as willed meaning, but this is just what premise 3 (a statement of intentionalism) asserts; thus, Iseminger assumes what he means to assert (in order to assert it). His ontological intentionalism does not prescribe a method for interpretation, but aims instead to tell us how things are (there is a notable anti-methodological trend in arguments for and against intentionalism since Hirsch). Hence his rather convoluted revision of the Identity Thesis (which is concerned with 'identity' in the sense of ontology rather than equivalence): 'A (typical) literary work is a textually embodied conceptual structure, whose conceptual component is (identical to) the structure – compatible with its text – which its author intended (meant) in composing it' (92).

Iseminger's formulation, despite its ontological subtlety, makes little advance on Hirsch in terms of its utility for the discipline of criticism. Ultimately, the work/text distinction is isomorphic with the speaker meaning/sentence meaning distinction discussed by Juhl, and others, and so tells us only that utterances are intentional, and should be read as such. It does not tell us, in what should be the first step of such an argument, why literary works should be read as utterances when, arguably, more might be derived from them in the absence of such a restrictive stipulation; the thesis of literary meaning as expression or communication is a kind of suppressed premise in Iseminger's argument. So the text/work distinction in itself offers scant grounds for his intentionalist thesis, particularly in light of the fact that authorial will/intention is not for him the only route to determinacy – he claims that 'indeterminacy *could* be resolved [. . .] by a random process' – but fails to elucidate his reasons for choosing authorial intention over some such 'random process' (84).

Although he denies that his conception of intention is explicitly mentalistic, yet his Hirschian-style claims about the indeterminate meaning of a particular sentence commit Iseminger to a view of the 'right' meaning as 'a function of some fact about the user', that is, 'some fact about what the person's thoughts were at the time he uttered the sentence (together with the fact that the sentence *can* be so used, of course)' (86). Ostensibly this mimics (with the addition of the parenthetic comment) the dual approach adopted by Hirsch, but Iseminger's emphasis upon his 'anticontextualism' (the claim that context does not determine meaning, although it may offer clues to it) reveals his leaning towards mentalism, the objections to which I have already touched upon (e.g., the openness to accusations of inaccessibility, irrelevance and/or idealism).

What Iseminger and Knapp and Michaels have in common is their dependence upon a view of intention as tied to a specifiable consciousness (the author's) and in this way neither account moves beyond (or, indeed, even manages to demonstrate the theoretical or methodological complexity of) Hirsch's own account. A fuller account of intention's relation to consciousness will be offered in Chapter 3, via a discussion of phenomenological theories of intentionality and reader-response theory; in the meantime it is worth noting that Richard Shusterman is alone among the latter-day intentionalists in maintaining that 'language is in some (but not necessarily mentalistic) sense intrinsically and irreducibly intentional' (1992: 66). It is in his claim that the intentional nature of linguistic meaning can be asserted without recourse to a reductive identification of textual meaning with authorial meaning that his advance appears to lie: 'The meaning of a text is inseparable from *some intention (or group of intentions) or another.* But the necessary meaning-securing intentions could belong to readers of the text rather than its original author' (67; my emphasis). This is a form of reader-response theory (or reader-response intentionalism, if you like). Shusterman's emphasis on the reader focuses on the experience of understanding (to which the reader has direct access) rather than on the experience of creation (which the reader must reconstruct or, in Hirschian terms, 're-cognise', falteringly) and so answers psychologistic criticisms regarding the inaccessibility of intention. Shusterman could, but does not, extend his 'intentional nature of language' argument to posit a more diffuse or spectral idea of agency. Instead he still wishes to tie intention to some consciousness or other (whether author's or reader's), a move that detracts from his bolder claim that language is intentional in a sense which is not necessarily mentalistic.

It is this idea – intentionality predicated of language and the utility of such a theory for our understanding of literary meaning – which I want to explore further, while keeping in mind the issues raised by this examination of Hirsch's ethical intentionalism. Namely, the vulnerability of an ethical argument to accusations of arbitrariness or subjectivism which suggests that an ethics of the text must be devised which is divorced from subjectivity but which yet retains some power for action in the world. This in turn indicates that the role of the author must be re-examined and perhaps replaced by a theory of intentional agency as a formal rather than a human quality. Meanwhile, the suspicion of origins and the objections raised against locating meaning in the mind of the artist which have together banished the author from much critical discourse, lead us towards a conception of intention as structure rather than genesis.[5] The (undoubtedly purposive) movement here is away from that blueprint in the mind of the artist which gave birth to the original fallacy. The next step will therefore be an appeal to speech act theory and speech act theories of literature, where intention is seen as central to the very process of meaning and understanding and where the opposition to be examined (and perhaps undone) is that between force and form. How successfully will a notion of intention as illocutionary force serve to banish the blueprint?

Notes

1 I don't believe, however, that Hirsch himself holds such an extreme position, although certain misreadings of his work have suggested that he does. His theory of linguistic conventions/types, which will be explored later in this chapter, allows for some linguistic meaning independent of intention; he simply disputes the relevance and/or utility of this in interpretation.

2 What Hirsch does not explore at this stage of the *Validity* argument is the animating power of the reader's intention; here it is the selecting/creating will, rather than the interpreting will which counts.

3 Of course, Knapp and Michaels, positing a much stronger version of intentionalism, would disagree – as will be discussed presently.

4 Curiously, Sidney Gendin, in his 1964 article on intentionalism, characterizes it as centred on the author's relation to the work, which is expressed either in the moral imperative that we should read the work according to the author's preferences, or in the empirical claim that the author is an authority on his own work; this second characterization, however, is much less common than Gendin seems to suggest.

5 To the extent that this opposition retains its usefulness – a usefulness which such a re-conceptualization of intention might ultimately throw into doubt by challenging the model of inside and outside upon which it is based.

Chapter 2

Intention, Illocution, Mimesis

Introduction: Speech act theory and literature

Another way of figuring intention in the literary context – distinct from those ways discussed in *Validity in Interpretation* and *The Aims of Interpretation* – is suggested by Hirsch during his participation in a conference on speech act theory and literature, the transcript of which was subsequently published in *Centrum* in 1975:

> After pondering performatives, [Austin] reached the more general and to literary students more interesting doctrine of illocutionary force. On this doctrine the meaning of a sentence is never *merely* the propositional meaning as determined by its words under the rules of the language. This locutionary or propositional meaning is *incomplete* until we know the unspoken intention – the illocutionary force behind the locution. (121; my emphasis)

Here he identifies 'unspoken intention' with 'illocutionary force' and argues that this is a vital addition to the bare or 'locutionary' meaning of the sentence, in order to make it properly meaningful. Hirsch is not alone in making this identification: Colin Lyas also explains 'the force of an utterance' as 'the purposes the speaker had in issuing the utterance', and in his rebuttal of Monroe Beardsley, Lyas blames him for '[basing] his attack on authorial reference too narrowly on facts about language that have to do with locutionary acts, or meaning in a narrow sense' (1986: 48, 46). The distinction being made here by both Hirsch and Lyas is one between (formal) meaning and use (force). The posited incompletion or 'narrowness' of locutionary meaning before the 'addition' of illocutionary force – which contextualizes and concretizes it – indicates that such intentionless or pre-intentional meaning is, at best, an abstraction, or at worst (as Knapp and Michaels would hold) impossible: not meaningful at all. Either way, a first step in the re-evaluation of intentional agency in literary production might be to examine applications of speech act theory to literary texts and, in particular, the conception of *illocutionary force* offered us by J. L. Austin, H. P. Grice, John Searle and others, to see whether this identification with intention is viable, and whether the basic distinction between form and force is one that holds.[1]

A speech act theory account of literature also has the advantage of allowing further consideration of those ontological issues unresolved by the preceding discussion of the New Critics and Hirsch, as it presents us with a vision of the text as socially and conventionally situated linguistic *act*, rather than (autonomous) linguistic *object*; this may help us account for the role of context and convention (in literary texts this encompasses issues of canonicity and genre, historical context, etc.) as they combine with intention in the production of meaning. This reconciliation of intention and convention proved taxing for Hirsch, as Chapter 1 detailed, although he faced the particular challenge (from which Austin exempts himself) of constructing an interpretative methodology: the 'generality' of Austin's theory, says Hirsch, is both its 'philosophical virtue' and 'a handicap to anyone who wants to use it as a methodological tool in criticism', but fortunately this is not my concern (1975: 123). For my purposes, another benefit of speech act theory is that it puts the onus on participation – on the vital relationship and shared preconceptions of speaker and hearer, author and readers – and thus moves some way towards an understanding of 'utterances', whether spoken or written, as intersubjectively constituted: 'acts' of 'speech' whose meaning is agreed upon publicly, collectively, rather than handed down from on high. It is the so-called tyranny of authors, their supposed right as 'oracles' to dictate and foreclose the meaning-possibilities of the text, that Wimsatt and Beardsley and the other anti-intentionalists were objecting to. By contrast, speech act theory presents us with an idea of Literature (with a capital L) as a social or collective practice, which accords with the way in which we, as a *community* of readers, experience it, and privileges this shared experience of meaning reconstruction over and above the author's experience of meaning 'creation' (with all the troublesome suggestions of authority, paternalism, singularity and individual genius which that term implies). It is also a useful corrective, should we desire one, to doctrines of semantic autonomy such as that proposed by Wimsatt and Beardsley.

Finally, the application of speech act theory to literary texts can facilitate a demystification of poetic language and literary production generally, if the conventions and characteristics inhering in so-called ordinary language (and speech act theory has its inception in the 'ordinary language philosophy' of the 1950s) are revealed to apply also to literature, indeed to all 'uses' of language. An account of literature as one among many possible 'uses' of language – rather than as some *thing* (or category of things) marked out by its ontological peculiarity, and decipherable only by those in receipt of special 'training' – permits us to see that 'use' as intentional, situated and convention-bound, as all uses of language are, and to see the determination of intention, context and the apposite conventions in each case as relatively unproblematic: a case of everyday, rather than specialist, interpretation.

If we employ Austin's terminology, it is the determination of illocutionary force that is central to understanding the utterances of others, and this requires us to make a decision about *how* or *in what way* speech is being used:

> When we perform a locutionary act, we use speech: but in what way precisely are we using it on this occasion? [. . .] It makes a great difference whether we were advising, or merely suggesting, or actually ordering, whether we were strictly promising or only announcing a vague intention, and so forth. (1962: 99)

The words themselves will not tell us this; if Hirsch is correct, the decision to be made by the interpretant concerns the 'unspoken' intentions accompanying an utterance, whether it is intended as advice, suggestion, order, promise or revelation of purpose. Unfortunately, the perceived 'peculiarity' of literary (and dramatic) uses of language causes Austin to exclude them from his account of performative utterances – or, at least, to proclaim that this is what he is doing – at the outset of *How to Do Things with Words* (1962):

> A performative utterance will, for example, be *in a peculiar way* hollow or void if said by an actor on the stage, or if introduced in a poem, or spoken in soliloquy [. . .]. Language in such circumstances is in special ways – intelligibly – used not seriously, but in ways *parasitic* upon its normal use – ways which fall under the doctrine of the *etiolations* of language. All this we are *excluding* from consideration. (22)

The proclamation 'we are *excluding*' is itself a performative utterance, although Austin's success in excluding literariness from the text of *How to Do Things with Words* is open to question. Certainly, the declaration of literature's non-seriousness has not prevented literary theorists such as Mary Louise Pratt and Sandy Petrey from attempting to outline accounts of the literary text as speech act; these accounts will be considered in due course. In the meantime, it is interesting to consider how the very distinctions which Austin invokes in the early stages of his lectures (constative/performative, locutionary/illocutionary/perlocutionary and ordinary/parasitic) come to unravel as those lectures proceed. This does not lessen the significance of his contribution to linguistic philosophy, but it does give licence for applications of this philosophy to literature, once the category of 'normal', non-etiolated discourse is revealed as possessed of fluctuating boundaries. In fact, Austin's own text is substantially infected by the literariness that he wishes to exclude. As evidence for his claim that *How to Do Things with Words* is 'a literary work through and through', J. Hillis Miller cites Austin's 'pervasive irony', his use of 'imaginary examples' and 'little fictional dialogues', all of which serve not only to illustrate and embellish, but also to significantly further his argument (2001: 40). Paradoxically, he uses this 'non-serious' language to convince us of the seriousness of his claims.

As can be seen from the example of the actor on the stage, and the somewhat indeterminate nature of his own lectures-cum-book,[2] it is not the fact of their being written down rather than spoken aloud which disqualifies certain (so-called non-serious) uses of speech from Austin's consideration, neither is orality itself a guarantee of inclusion; so what prejudice there is here concerning literariness is not an instance of phonocentrism, as Derrida claims (1982: 321ff.); Austin uses the term 'utterance' flexibly enough to include instances of writing as well as speech. It is the very *theatricality* of poems, soliloquies and so on which marks them out as marginal, suggesting that there are good and bad forms of performativity, as far as Austin is concerned. Nevertheless, his key recognition is that language does more than just report facts or reflect reality in a manner that is subsequently verifiable, that utterances also *indicate* or *perform* certain actions not necessarily bound up with the making of statements. This also would seem to open the door to the consideration of literary texts, as the ordinary language/poetic language distinction customarily rests on a view of literature as 'peculiar' due to the supposed suspension of reference that occurs with fictional statements, their disjunction from the 'real world' and from truth values.

Austin is here trying to move on from an account of language as representational or descriptive or denotative, towards a more sophisticated analysis of what and how language (en)acts (things) in the world and literary utterances may be said to 'act' in myriad ways, not all of them distinguishable from the actions performed by so-called ordinary speech. In addition, Austin troubles the values of truth and falsity, as applied to utterances, by relativizing them:

> It is essential to realise that 'true' and 'false', like 'free' and 'unfree', do not stand for anything simple at all; but only for a general dimension of being a right or proper thing to say as opposed to a wrong thing, in these circumstances, to this audience, for these purposes and with these intentions. (1962: 145)

Truth and falsity, then, are replaced by more nebulous notions of 'propriety' – such as successful/unsuccessful, felicitous/infelicitous, happy/unhappy – all of which can perhaps be more satisfactorily applied to literary texts (and critical readings of those texts) than the terms 'true' and 'false'. Furthermore, the 'happiness' or 'propriety' of a particular utterance is something agreed upon, that is, conventional and intersubjective; again this is reminiscent of literary critical discourses on the 'success' of a work *as a novel* or *as a poem*. The criteria for success in literary works are genre-specific, historically specific – because tastes and ideas about literature (what it is and does) change – and, generally, agreed upon by communities of readers and critics, therefore similarly intersubjective.

Performative-constative and the intentionality of use

In his preliminary – and admittedly provisional – 'isolation of the performative', Austin contrasts: utterances which 'describe' or 'report' or 'constate' something, and which can, accordingly, be judged as true or false (i.e., constatives); and utterances which do not do this and where 'the uttering of the sentence is, or is a part of, the doing of an action, which [. . .] would not *normally* be described as, or as "just", saying something' (i.e., performatives) (1962: 4, 5). The contrast here is between speech that 'says' and speech that 'does', but what Austin discovers in the course of the lectures is the impossibility of isolating either the constative or the performative, because all speech can be seen as performing some sort of action (even if that action is merely 'stating') and all utterances are best judged in terms of 'happiness' and 'unhappiness' rather than truth and falsity. The gradual dissolution of this distinction, and Austin's realization that performativity infects all utterances to some degree or other, also bears upon the closely analogous locutionary/illocutionary distinction. It is this latter distinction which is so vital to those intentionalists wishing to identify intention and illocution and so demonstrate the inextricability, but non-identity, of meaning and intention which would therefore make intention (in its own right) something worth appealing to. Of illocutionary and locutionary acts he writes:

> These two kinds of acts seem to be the very things which we tried to use, under the names of 'doing' and 'saying', as a means of distinguishing performatives from constatives. If we are in general always doing both things, how can our distinction survive? (133)

The answer is that it can survive only in the abstract, as Austin recognizes. So, to apprehend the notion of the constative, we must 'concentrate on the locutionary', but in order to exclude 'the illocutionary aspect' we are thereby made dependent upon 'an over-simplified notion of correspondence with the facts', which presumes that such 'facts' are universal and such correspondence fixed for all time, for everyone, in all circumstances and so on (146). This sits uneasily alongside Austin's earlier dismissal of the truth/falsity axis in favour of more relativistic values.

The performative, too, in pure conceptual terms, requires attendance to illocutionary force alone, abstracted 'from the dimension of correspondence with facts', yet the very fact that our speech is able to act at all shows us that it must stand in some sort of efficacious relationship to some sort of world (although, of course, this need not be a relationship of correspondence) (ibid.). Austin's doubt is clearly manifested – 'perhaps neither of these abstractions is so very expedient' he confesses – but it is only as abstractions that his linguistic categories can survive (ibid.). He confesses that 'in general the locutionary act as much as the illocutionary is an abstraction only: every genuine speech-act is both' (147). This could be read, by those with an interest in doing so, as chiming with Knapp and Michaels' dismissal of 'intentionless' meaning and Stanley Fish's description of a sentence without illocutionary force – that is, a 'bare' locution – as 'just a series of noises, a dead letter with no more "content" than a list of words' (1980: 229). Hirsch is, then, perhaps wrong to represent intention as an illocutionary 'addition', because even Austin apparently comes to the conclusion that the locutionary and the illocutionary cannot be separated: so *there is no locution* prior to the 'addition' of 'force', and force is, in fact, not an addition at all, but always already there in utterances (although not, perhaps, in lists, for instance).

By extension, the separation of meaning and intention (if configured as form and force, or locutionary and illocutionary, respectively, as I am suggesting here) can be seen as premised upon some idea of them as abstract, isolatable concepts – the only way in which they can be conceptualized as distinct. Once we are in the realm of speech (rather than dealing with language in itself) and speech acts – that is, actual, contextualized, singular instances of speech – the distinction breaks down, and each utterance is meaningful through a combination of form and force, every genuine speech act is both locutionary and illocutionary; we cannot think one without the other, and one is not prior to the other. How is such a conclusion useful for the beleaguered intentionalist? On the one hand it situates intention at the heart of the process of meaning and understanding, which may seem to militate against its forced banishment; on the other, the very impossibility of separating meaning and intention may appear to obviate the need for a definite appeal to the latter, if all utterances (i.e., all *uses* of language) are always, inescapably, intentional; what has yet to be successfully argued is that all literary texts may profitably be viewed as utterances in the first place. As for the question of exactly *whose* (or, more speculatively, *which*) intention or intended force is in play, this is open to debate. In J. Hillis Miller's reading of Austin, he appears to place the emphasis upon the reader, that is, on how the words are *taken* rather than on how they are *used*, thus meaning 'is not a matter of some objective feature of the words themselves but a matter of how any given set of words is taken' (2001: 17). This is an unacknowledged deviation from Austin's text, although Miller is right to suggest that neither 'the words themselves' nor the way in which they are read/understood has any kind of 'objective' status; a point with which I think Austin would concur.

This declared absence of locatable, objective meaning is the reason why it is use, rather than meaning, that the speech act theorist is attempting to isolate: rather than asking what an utterance *means* (in this context, in particular circumstances, etc.), he asks in what way an utterance is being *used* (to warn, to suggest, to promise, e.g., in these circumstances, etc.). The suggestion is that use, because it is conventional and public, is accessible in a way that meaning – which yet remains an underdetermined notion – is not; although Austin, always amiably keen to undermine his own arguments, concedes that the class of illocutionary acts is not exhaustively determinable, thanks to the 'different senses of so vague an expression as "in what way are we using it"' (1962: 99). In addition, and despite the shift in focus from meaning to use, from locutionary to illocutionary, the issue of intention remains central to the process of understanding in what way utterances: are used; 'we need', says Austin:

> to establish with respect to each kind of illocutionary act [. . .] what if any is the specific way in which they are intended, first to be in order or not in order, and second, to be 'right' or 'wrong'; what terms of appraisal and disappraisal are used for each and what they mean. (146–47)

To judge the 'felicity' or 'happiness' of an illocutionary act (a category which may now be seen to include all speech acts) we must know in what way it is intended to be felicitous or non-felicitous: it is intention which determines the criteria of felicity. In many ways, however, this is as troublesome as the claims of authorial intentionalists, because it seems to appeal to something outside the act or utterance itself, either to some higher authority or to some inward (i.e., psychological) essence or state – in this case, that of the speaker. This is at odds with Austin's avowed anti-mentalism, but his choice of verbs 'in the first person singular present indicative active', that is, 'explicit performatives' such as 'I do' or 'I bet you sixpence', to form the cornerstone of his account, means that the issue of sincerity has to be addressed (5).

Sincerity, it seems to me, is central to the discussion of speaker or author intention, because being sincere involves not merely 'saying what you mean' – indeed you may make a mistake, use a word wrongly, and still be 'sincere', your integrity remaining intact – but meaning (i.e., intending) what you think you are saying; it is a matter of psychological make-up and good faith rather than achieved effect. This is what makes sincerity problematic: how can we legislate for the mental states of others? Is a reading of 'outward' behaviour always sufficient to establish sincerity or insincerity? In fact, an uneasy plea for sincerity is written into Austin's 'conditions for happy performatives', where he attests that:

> Where, as often, the procedure is designed for use by persons having certain thoughts or feelings, or for the inauguration of certain consequential conduct on the part of any participant, then a person participating in and so invoking the procedure *must in fact have those thoughts* or feelings, and the participants must intend to so conduct themselves. (15; my emphasis)

So despite the stress laid on public, social, conventional procedures – procedures which can be legislated for, and which are by nature rule-bound – Austin still finds himself dependent on vague imprecations to the effect that the speaker must 'mean' what he says, otherwise the utterance is potentially 'unhappy' or void. Nevertheless – and I think Stanley Fish is wrong to claim that 'the question of what is going on inside, the question of the "inward performance" is simply bypassed; speech-act theory does not rule on it' (1980: 203) – still it is clear that Austin does not want to make intention the decisive factor in the process of meaning and understanding, as he earlier warns against the view that the 'seriousness' of utterances:

> consists in their being uttered as (merely) the outward and visible sign, for convenience or other record or for information, of an inward or spiritual act: from which it is but a short step to go on to believe or to assume without realizing that for many purposes the outward utterance is a description, *true or false*, of the occurrence of the inward performance. (1962: 9)

In this way he argues against the interpretation of utterances as 'merely' the visible, public manifestation of some mysterious, irretrievable prior blueprint. There is, nevertheless, a vacillation in *How to Do Things with Words* between the ratification of, respectively, 'inward' and 'outward' acts (where intention is conceptualized as definitively 'inward', and so potentially inaccessible). What is the relative hierarchical standing of these acts? Notably, he suggests that their 'inwardness' or 'outwardness' is delineable and fixed; no blurring of the boundaries occurs; this is analogous to Wimsatt and Beardsley's claims regarding the definition of what is internal and what external to the text.

Crucially, Austin prevaricates over which act, inward or outward, will be decisive in determining the level of responsibility incurred by a speaker. It is the transformation of utterances into acts which introduces into Austin's account the question of responsibility, an apparently natural concomitant of agency: we are tied to our actions, even our verbal actions, and must therefore be held accountable for them and, to a certain extent, for their consequences. Paradoxically, it is this ethical angle to his linguistic philosophy which necessitates the evidently undesirable (for Austin) discussion of sincerity. Placing the emphasis on the external act, as he on the whole attempts to do, constitutes an endorsement of the policy that 'our word is our bond' and illustrates the moral obligations that accompany such acts of speech: 'Accuracy and morality alike are on the side of the plain saying that *our word is our bond*' (1962: 10). In other words, the speaker must not be allowed to disclaim all responsibility with the convenient cry of 'I didn't mean it', when reneging on a promise or bet – or marriage. Miller detects an ideological imperative here, namely, a desire for the imposition of order, both legal and linguistic (suggesting, I suppose, that accuracy *equals* morality for Austin):

> The performative must *not* depend on the intentions or sincerity of the one who speaks. If Austin's theory is to be cogent, and if he is to attain his goal of securing law and order, the words themselves must do the work, not the secret

> intentions of the speaker or writer. For civil order to be maintained, we must be able to hold speakers and writers responsible for their words, whatever their intentions at the time. (2001: 29)

And sincerity is not something that can be legislated for, if we persist (and both Austin and Miller do appear to persist) in thinking of intention as something 'secret', rather than as something that can be read through or in the 'outward' action, or as something that the action itself demonstrates or displays. Fish, as it happens, does suggest this, in arguing that one consequence of Austin's failure to rule on 'what is going on inside' (a claim that I have disputed) is that intention becomes something 'outward' or at least outwardly appropriable:

> Intention, in the view of [speech act theory], is a matter of what one takes responsibility for by performing certain conventional (speech) acts [. . .] intentions are [as a consequence of the neglect of 'inward performance'] available to anyone who invokes the proper (publicly known and agreed upon) procedures, and [. . .] anyone who invokes those procedures (knowing that they will be recognised as such) takes responsibility for having that intention. (1980: 203–04)

Therefore, uttering certain words in certain circumstances commits me to having certain intentions: the intention to get married, agree on a bet, name a ship or make a promise. Intention is, according to Fish's admirable pragmatic demystification of it, something that we have access to insofar as we have access to various, possibly culturally specific, conventions.

Yet the English judicial system – largely responsible for maintaining the 'civil order' of which Miller writes – in fact places a strong emphasis on intent which *may not* be revealed in the action or the conventions surrounding it, particularly when distinguishing between instances of murder and manslaughter: the outward act of killing, even when proven, is not itself sufficient to determine the verdict. Austin too, is forced to recognize that saying alone is not enough 'to make it so': the right conditions must obtain, and these include (without exclusively consisting in) the requisite intentions for the accomplishment of the act (intending alone is certainly not enough 'to make it so'). Indeed, it is largely on the basis of such a sincerity condition that Austin is able to exclude from his account so-called literary uses of language. It is the different intentions lying behind such utterances which exclude them: for example, the desire to entertain, to contribute to plot or narrative flow, to create an imaginative world, rather than to effect changes or practical consequences in the real world. This view of literature as radically divorced from the real is, of course, highly disputable; nevertheless, literary utterances are perceived as 'insincere', in a special sense of that term. Without this sincerity clause for excluding non-serious usage,

Austin seems to think that he would be unable to explain why saying 'I do' or 'I promise' as part of a play did not constitute some kind of morally binding obligation. Yet, the (other) conditional requirements of 'an accepted conventional procedure' and the presence of all and only those 'particular persons and circumstances [. . .] appropriate for the invocation of the particular procedure invoked' would seem to be sufficient to exclude poems and dramatic soliloquies without having to legislate for the recalcitrant mental states of the speaker (1962: 14, 15).

The example – again, generously proffered by Austin himself – which could, however, problematize the serious/non-serious distinction is that of the joke, as the external trappings of such an utterance, including the words used, could be indistinguishable from a 'serious' utterance. What, apart from intention, will allow us to discriminate between two outwardly identical speech acts, one serious, the other uttered in jest? Context, possibly, but satirical intent has been overlooked in written and spoken utterances alike, either because the context is not revelatory or because it has been misread, and in fact the comedy frequently arises out of this very misreading or at least out of the arguably indeterminate nature of the utterance. Ultimately, Austin's own delightful humour (which is surely intentional) consistently serves to undermine the imposition of order (whether in the form of clear-cut linguistic categories or workable social structures) which appears to be the central intention of his text – proving, perhaps, that an author's own intentions will often work against him and against each other.

Intention and reason: Talking as purposive behaviour

Unlike Austin, H. P. Grice has no qualms about placing intention at the centre of his theory of speech acts and his particular conceptualization of intention in a speech context is notable in the way it ostensibly moves away from the unsatisfactory 'blueprint' model. Although he claims that there may be instances 'where an utterance is accompanied or preceded by a conscious "plan", or explicit formulation of intention', yet, he says, 'explicitly formulated linguistic (or quasi-linguistic) intentions are no doubt comparatively rare' (1989a: 222). In a series of articles published between 1957 ('Meaning') and 1989 ('Meaning Revisited'), he argues that understanding meaning *just is* understanding the utterer's intention. As Max Black recognized, in his consideration of Grice, intention serves here to mark 'a sharp distinction between human *action* and mere animal behaviour' (1973: 258). The use of 'mere' is revealing: as the discussion of Hirsch's authorial intentionalism revealed, an ethical imperative, in the form of a tacit humanism, is often to be found at the base of intentionalist theories of communication. What Grice is seeking to bring out, then, is the definitively human nature of speech and understanding.

According to Grice's theory of non-natural meaning, grasping intention is not a separate or prior act but integral to the process of understanding.[3] Indeed, for Grice, successful understanding *only* occurs with this recognition of intention: for a person to 'mean' something by X (an utterance) he 'must intend to induce by X a belief in an audience, and he must also intend his utterance to be recognised as so intended' (1989a: 216). The reflexivity of this account indicates that there are two key intentions at work here: intention to mean and intention to be understood as intending to mean. These are not identical, but neither are they independent of each other: 'the recognition is intended by A [the speaker] to play its part in inducing the belief', claims Grice (219). Such intentions serve to delineate the speaker or author's expectations of his audience, the contract agreed between them. So then, recognition of intention is built into the account as a necessary feature, such that it is not sufficient for the audience to grasp the intended meaning, they must also grasp it *as intentional.* As a means of illuminating its logical structure, the Gricean account can be written schematically, as follows:

1. the speaker intends to produce a particular *response* in the hearer;
2. the speaker intends the hearer to *recognize* intention 1;
3. the speaker further intends this recognition (2) to form (at least) part of the hearer's *reason* for their response – in other words, the hearer's response cannot be contingently that intended by the speaker, but must be at least partly caused by the hearer's recognition of the speaker's intentions, etc.

One suppressed premise in Grice's account is the requirement that the speaker must believe such recognition of intention by the hearer is at least possible (and even likely); this is reminiscent of his claim elsewhere that 'one cannot have intentions to achieve results which one sees no chance of achieving' (1969: 158) – and the speaker also assumes here that it is necessary to induce the desired belief in the hearer this way, that is, that they would probably not come to hold this belief except by means of this recognition of intention.

This insistence on recognition of intention, on the non-accidental attainment of understanding, reveals quite how dependent on intention Grice's account is. Why does he exclude as aberrant cases in which the hearer successfully comes to hold the desired belief without, however, recognizing that this is what the speaker wants (intends) his hearer to do? In other words, cases in which 1 (response) is achieved, but 2 (recognition) and 3 (reason) are not; or, more specifically, cases where 1 is achieved but not *by means of* 2 and 3. What is it about cases in which only the first intention (to provoke a particular response) is fulfilled that disqualifies them as instances of successful communication? Grice appears to be making a point here about the importance of recognizing meaning as intentional: he is not quite saying that meaning without recognition of intention is not meaning, but he is saying that a (contingently) correct response, without the recognition that this is the response intended by the speaker and the marshalling of this recognition as a reason for that response, is not in fact a 'correct' response at all. By comparison with Austin's largely conventional conditions for what makes a successful speech act, Grice's conditions as stated here are overwhelmingly 'internal' (to the particular, singular act of meaning and understanding) and mentalistic. Paradoxically, the very emphasis on participation between speaker and hearer which speech act theories insist upon, leads Grice to legislate for the beliefs and systems of reasoning of his speech situation protagonists in order to ensure this participation at the highest level.

On the other hand, the particular emphasis on the production of a response in the hearer (by means of his recognition of the speaker's intention, etc.) suggests that Grice's concerns are in fact predominantly *perlocutionary*. The initial description of 'inducing a belief' is subsequently generalized to '[producing] some effect' but both expressions indicate a focus on something outside or beyond the utterance itself, even if the effect is something that occurs in the mind of the hearer, such as getting them to believe something (1989a: 220). Is the production of an effect to be achieved in addition to (i.e., separately from) the comprehension of the meaning, or is the belief to be induced merely a belief about what the speaker means on this occasion? Grice is unclear on this, unhelpfully averring that 'to ask what A meant is to ask for a specification of the intended effect' (ibid.). Yet meaning and effect are quite obviously distinct, particularly if the theory is extended to cover textual utterances: a reader may comprehend the meaning(s) of a text while reacting in a way (e.g., with hostility or disagreement or disbelief) neither intended nor welcomed by the author. It is not hard to devise hypothetical speech situations in which a similar situation might obtain. In addition, we may ask whether the 'speaker' of an utterance – whether that 'utterance' is spoken or written – is always trying to produce some singular, determinable effect. In literary texts in particular, the author's refusal to set definite parameters for the possible responses of the reader (the text's effects upon that reader) is not generally seen as a fault. Grice's overly proscriptive account – suggesting that the author can or should delimit not merely the illocutionary but also the perlocutionary force of the utterance – contradicts the traditional speech act theory emphasis on the collective production and/or determination of meaning. Although Grice qualifies his account with the claim that only the speaker's 'primary intention' – that is, the main intended effect, not any and all possible further effects (intended or otherwise) which may follow from this – is relevant to an utterance's meaning, still it seems that the author-speaker as sole source of meaning has crept back into the picture (1989a: 221). The very designation of (an) authorial intention as 'primary' is suggestive of its pre-eminence and determinative power. As is becoming apparent, this didacticism (a conception of absolute *author*ity in communication) is symptomatic of theories of meaning which place authorial or speaker's intention at their centre, even if this is then tempered by an emphasis on convention, context and co-operation.

There is a notable distinction made in the Gricean schema between reasons and causes: it is not enough that the hearer's response be *caused* by their recognition of the speaker's intention, in order to preclude this situation obtaining accidentally, they must (consciously or otherwise) cite the intention as part of their *reason* for responding in that way. In other words, parts 1 and 2 of the schema are interlinked in that 2 (recognition) must provide the reason for 1 (response). This distinction serves to reinforce that opposition of *human action* and *animal behaviour*, previously mentioned, and, more generally, a distinction between the intentional and the natural; the participants' rationality is both assumed and necessary. As Black realizes, in his analysis of Grice, 'the chief enemies of an intentionalist theory of meaning [. . .] are *causal* theories, such as those of Charles Morris or Charles Stevenson, which offer a "naturalistic" or behaviouristic account of meaning' (1973: 258). The threat such theories pose to intentionalism lies in their exclusion of some normative model of rationality, perceived as the prime indicator of humanity; it is instructive to consider how an idea of reasonableness structures Grice's theory of meaning, setting limits on what and how a speaker may mean. This is evident in even his more general pronouncements about communication – witness his assertion that 'our talk exchanges do not normally consist of a succession of disconnected remarks, and would not be rational if they did' (1989b: 26); the implication here is that they would also not be intelligible. Yet literary 'utterances' may and do use disconnection and disorder to good effect, and the presence of 'literariness' in 'everyday' speech (or speech that sets itself up as 'normal', such as Austin's in *How to Do Things with Words*) suggests that not all utterances are bound by this requirement to be rational.

Such a requirement is more formally enshrined in Grice's 'Cooperative Principle', developed to delimit what conversational 'implicature' is permitted. To this end, he produces a series of conversational 'maxims', which are divided according to category:

1. Quantity (be as informative, but not more informative than is required)
2. Quality (an invocation to truth: don't say what you believe to be false or that for which you lack adequate evidence)
3. Relation (be relevant)
4. Manner (an invocation to perspicuity: avoid obscurity, ambiguity, unnecessary prolixity, disorder) (1989b: 26–27).

To abide by these is to abide by the 'Cooperative Principle' and to accede to Grice's assessment of 'talking' as 'a special case or variety of purposive, indeed rational, behaviour' (28). Awkwardly, Grice maintains both that we *should* follow the Cooperative Principle, and that this is just what we *do* actually do in conversation: 'it is just a well-recognised empirical fact that people do behave in these ways; they learned to do so in childhood and have not lost the habit of doing so' (29). Therefore, if we assume that a speaker is, as expected, adhering to the Cooperative Principle (its injunctions to brevity, relevance, perspicacity, etc.), then we are able to read certain implications in and of the speaker's utterances as valid, and disqualify certain others as invalid.

Again, though, literary texts are evidently not required to be informative, truthful or relevant, and may profitably exploit obscurity, prolixity and so on as worthwhile and effective literary devices. Grice recognizes this and so allows for the (crucially) intentional 'flouting' of the perspicuity maxim, for example, in cases of poetic ambiguity: 'There seems to be no alternative to supposing that the ambiguities are *deliberate* and that the poet is conveying *both* what he would be saying if one interpretation were intended rather than the other, and vice versa' (35; my emphasis). But ambiguity in literature (which of course is not the only place in which it occurs or is troublesome) often brings into play more than two clear-cut and simplistically antithetical or contradictory meanings, as Grice imagines here. Poets, then, are permitted to 'opt out' of the Cooperative Principle and in such cases the speaker makes it clear that the maxims do not obtain, and that this is an intentional flouting, hence the 'deliberate' nature of the ambiguity in Grice's hypothetical poem. Literature may, perhaps, be seen as having its own set of 'maxims' governing implicature, maxims agreed between author and reader, although it seems unlikely (given Hirsch's failure to do so) that a definitive, universally applicable set of these could ever be specified; that is, a set particular enough to be worth having (e.g., beyond the 'agreement' that a character or event is fictional) and general enough to hold across different literary contexts, bearing in mind the variety of 'things' and 'objects' which may be grouped under the heading of 'Literature'. The apparent inapplicability of the Cooperative Principle to literature arises out of our intuitive assumption that literary utterances constitute *more* than 'a maximally effective exchange of information' – Grice concedes that this remit is 'too narrow' a specification even for conversational exchanges – and so it is (28). An awareness of the impossibility of maintaining this opposition of literary and ordinary/conversational language (an issue which Grice does not address, here or elsewhere) would trouble his definition of communication as (merely) the exchange of informational content still further.

The casual linking of purposiveness and rationality in Grice's central thesis about the nature of 'talking' reveals how crucial Enlightenment-inspired notions of reason are to intentionalist theories of meaning: intention is always the right sort (i.e., the most 'reasonable' sort) of directedness and communication, as a result, is always *intent upon* (i.e., directed towards) clarity and truth. The development of a sound interpretative methodology appears to necessitate this agreement of rationality between utterer and interpretant, speaker and hearer, which requires that each acknowledges and respects the particular 'direction' of the 'conversation' and agrees that there is some thing (usually, the truth) towards which that instance of communication will tend. It is in this way that communicative exchanges are seen to be co-operative:

> Each participant recognizes in them, to some extent, a common purpose or set of purposes, or at least a mutually accepted *direction*. This *purpose or direction* may be fixed from the start (e.g. by an initial proposal of a question for discussion), or it may evolve during the exchange; it may be fairly definite, or it may be so indefinite as to leave very considerable latitude to the participants (as in a casual conversation). (26; my emphasis)

However 'indefinite' or unfixed the 'direction' of the exchange, yet its 'evolution' must be along rational lines if it is to be more than a succession of 'disconnected', and so meaningless, remarks. The suggestion here that a conversational exchange itself displays a discernible 'purpose or direction'[4] seems particularly useful both for recognizing the transactional (intersubjective) nature of such an exchange and for shifting the locus of intention away from individual subjects and towards the utterance itself. Nevertheless, the purposiveness of which Grice writes here is definitively uni-directional, teleological and to a large extent initiated and controlled by the speaking subject. Moving away from this uni-directional account of purposiveness in linguistic exchanges will be one of the central challenges of my final chapter on formal or textual intentionality.

Mimesis and make-believe: Literature and pretended speech acts

One way in which Grice retreats from acknowledging the violation of his Cooperative Principle by literary utterances is by claiming that 'the poet is not explicitly saying any one of these [apparently contradictory] things [that he appears to be stating] but only conveying or suggesting them' (1989b: 35–36). In other words, the propositional force of the literary statement is diminished (or even denied absolutely), because the poet is not so much 'using' language as 'mentioning' it. This suggestion that literary texts only represent or pretend to perform speech acts, rather than actually performing them or constituting actual speech acts themselves, receives a more explicit formulation in the work of Monroe Beardsley, John Searle and Richard Ohmann, among others. In fact, the view of literature as (merely) mimesis – a peculiar and straitened kind of agency when set alongside the assertive power of 'real' speech acts – is notably pervasive within the field of literary aesthetics.

Richard Ohmann is the most persistent purveyor of a mimetic theory of literary speech acts.[5] The following is the most succinct, and most frequently cited, statement of his position: 'The *illocutionary force* [of a literary work] is mimetic. By "mimetic", I mean purportedly imitative. Specifically, a literary work *purportedly imitates* (or reports) a series of speech acts, which in fact have no other existence' (1971b: 14). But there is room for ambiguity even here: does the author genuinely 'imitate' speech acts, or only 'purport' to imitate? Or does 'purportedly' refer to the fact that an author of fiction 'purports' to make speech acts, but does not in fact do so? This is Beardsley's contention, that fictive discourse is 'discourse in which there is a make-believe illocutionary action, but in fact no such action is performed' (1978a: 170). The reason 'no such action is performed' is that 'one or more of the requisite conditions are lacking' but the reader is nevertheless asked to participate in the process of 'make-believe' by imagining that what is lacking is in fact present. What literary-fictional discourse is perceived to lack is illocutionary force – the force which will make it an action of one sort or another; the shared pretence that illocutionary force is present is what turns 'a genuine illocutionary action' into 'a fictive one' (which, presumably, is not 'genuine') (ibid.). Yet, in the light of the preceding discussion, it is difficult to explain: first, just what sort or sorts of actions a literary text does perform, above and beyond the action of 'pretence' or 'mimesis'; and second, how it is that literary discourse is meaningful at all, given the posited inseparability of the locutionary and the illocutionary. Is there, then, some special class of illocutions which includes 'the illocutionary act of telling a story or writing a novel' (Searle 1979: 63)? Searle thinks not, but he does set out to discover how the words which constitute a literary text can be meaningful, when apparently stripped of their illocutionary potential: the form remains the same (we use the same words and sentences in fiction that we can and do use in 'ordinary' speech), it is the force which is altered (these words do not have the power to assert, question, promise, etc.). So the mimesis argument posits some kind of separation of form and force; this suggests that form without force can be meaningful, and that illocutionary force is not, therefore, identical to meaning – or to intended author meaning, as Hirsch tried to claim.

The key tenet of Searle's argument is his emphasis on the rule-governed nature of all genuine speech acts.[6] For Searle, 'speaking or writing in a language consists in performing speech acts of a quite specific kind called "illocutionary acts"' (58). So far, so uncontroversial (the identifying of 'speech act' and 'illocutionary act' is widespread) – and, furthermore there exists 'a systematic set of relationships between the meanings of the words and sentences we utter and the illocutionary acts we perform in the utterance of those words and sentences' (ibid.). What is important here is that the relationship between meaning and force or illocution is *systematic*, that is to say: governed by rules and patterns which are conventional and discoverable. Interestingly, Searle says nothing about the possible arbitrariness of those rules or the occasions (not only within fictional discourse) on which they might be subverted or violated.

The anomalous nature of fictional discourse – and Searle is explicitly writing about the category of 'fiction' rather than 'literature' in general – arises from the suspension of these rules, or their substitution by some as yet unspecified other set of rules and conventions: 'in fictional speech semantic rules are altered or suspended in some way' (60). Indeed, it is precisely the suspension of these rules which alerts us to the fact that what we are dealing with is fiction rather than non-fiction, according to Searle; fictionality is therefore an internal property of a piece of discourse, but is it a discoverable property? Searle does not concede the epistemological circularity of his argument here: a text is known to be fictional because it merely pretends illocutionary force, and a text is known to merely pretend illocutionary force because it is fictional. A text's fictionality may make a difference to our reading and understanding of it, but how do we first come to register this difference? Why, says Searle, because the author tells us so (at least we hope he does): whether or not a work is fiction 'is for the author to decide' (59).

By contrast, Monroe Beardsley, although conceding that 'a literary work is something produced intentionally' – he could hardly claim otherwise, unless he wants to pursue the 'monkeys with typewriters' line of reasoning – persists in his belief in the fallaciousness of looking to authorial intention to determine fictionality (or anything else), arguing that: 'considerations about intention are not logically requisite to a decision whether [a certain text] is a poem', and that such considerations result in a circular definition of literature (1978a: 165, 168). This leads him to the frankly uneasy postulation of such a thing as an 'aesthetic intention' in the place of any kind of 'intention to write a poem' (168). Searle takes the latter as an apparently less contentious bedrock for his argument and admittedly does not really conceal his leanings towards a version of authorial intentionalism, subsequently writing of the intentional fallacy:

> Perhaps there is some level of intention at which this extraordinary view is plausible; perhaps one should not consider an author's ulterior motives when analyzing his work, but at the most basic level it is absurd to suppose a critic can completely ignore the intentions of the author, since even so much as to identify a text as a novel, a poem, or even as a text is already to make a claim about the author's intentions. (1979: 66)

And the intentions that matter when it comes to identifying a work as fictional or non-fictional are, notably, 'complex illocutionary intentions', as 'there is no textual property, syntactical or semantic, that will identify a text as a work of fiction' (66, 65). Illocutionary intentions – despite their power to dictate fictionality (which is, remember, an 'internal' property of a text) – are firmly situated outside and prior to the text. Searle does not allow for the realization of intention in or through the textual, syntactic and semantic properties that he makes reference to here, but instead limits intention to decisions made at the moment of creation (whenever this moment might be, whatever its duration).

The troubling of the usual word-world relation (and the rules which govern this) which Searle claims takes place in fictional discourse, can be seen as analogous to the intentional 'flouting' of Grice's Cooperative Principle. As with Grice's hypothetical piece of poetry, fictional discourse for Searle is not a 'violation' of the rules (as lying is) but constitutes an entirely 'separate language game' from ordinary speech (1979: 67); which particular language game is being played is determined by which language game the author or speaker is intending to play. Searle does not consider the possible consequences if the hearer or reader disagrees about which game in particular is being played. Again, the abrogation of these semantic rules (which amounts to the rescission of illocutionary force) has ramifications for the responsibilities and commitments which attend any kind of utterance, such that 'if the author of a novel tells us that it is raining outside he isn't *seriously* committed to the view that it is at the time of writing actually raining outside' (60; my emphasis). The writer, then, is not required accurately to represent the world. The assumed relationship of correspondence between word and world (language and reality both figured here as incontestably stable entities) breaks down, because 'what makes fiction possible [. . .] is a set of extralinguistic, nonsemantic conventions that break the connection between words and the world' (66). However, being 'nonsemantic', the conventions which facilitate fiction do not break the *semantic* connection; if they did so, the words which make up fictional discourse would not be meaningful. By contrast with the 'vertical' rules which govern non-fictional discourses, those applicable to fiction are 'horizontal' (ibid.). 'Horizontal' suggests, perhaps, that they take in considerations of genre (the identification of a work as being of one genre or another, and its fidelity to that genre) and of the internal coherence (the relation of parts to whole) of the work, rather than constituting rules of reference, tying the discourse to its referents, to something outside itself; but such reasoning posits the literary text as a peculiarly free-floating entity.

The paucity of 'serious' commitments invoked by fictional utterances (the most serious of all being, naturally enough, a commitment to truth) also signals Searle's debt to his philosophical predecessor, Austin, at least in his choice of terminology. In distinguishing between 'fictional' and 'figurative' speech, Searle elects to call metaphorical uses of language 'nonliteral' and fictional uses 'nonserious' (ibid.). The 'nonseriousness' of fictional discourse is such that what appear to be fictional assertions are in fact not assertions at all, as they are not bound by, for example, the 'essential rule', which entails a commitment to truth, and the 'sincerity rule', which commits the speaker to a belief in that truth (62). Both are 'vertical' rules, in the way that they tie the assertion to something outside itself, respectively a state or feature of the world, and the thoughts or mental state of the speaker/writer at the time of speaking; both denote responsibilities which the speaker is assumed to agree to, as a matter of inevitability, in the act of speaking. The writer of fiction, however, is not bound by such 'commitments', neither is he required to be suitably knowledgeable to make the assertion, nor is the assertion itself required to be worthy of communication ('worthiness' being determined by whether or not the informational content is already known and, indeed, whether it is worth knowing and communicating).

From both Grice and Searle, then, we get a highly idealized account of communication as maximally efficient, worthwhile, sincere and intent upon truth; neither appears to recognize the impossibility of establishing beyond doubt whether an utterance or speaker is successfully fulfilling these actually quite relativistic conditions. How do we test for sincerity if it is a matter of the author or speaker having the 'right' thoughts? The only evidence we have for the presence of these thoughts is the utterance itself, yet we cannot determine what kind of utterance this is (sincere, insincere, fictional or non-fictional) until we know if the sincerity condition has been met. Nevertheless, it is compliance with (at least) these rules which makes an utterance an assertion by way of making it also, significantly, an illocutionary act. The wilful *non-compliance* of fictional utterances means, therefore, that they cannot be classed as illocutionary acts; this is Searle's logic.

But the non-compliance is at least *wilful*. For Searle, intention is central to the process of mimesis, the special kind of non-deceptive pretence (a dramatic representation of illocutionary force which is only that: a representation) which occurs in the production of fictional discourse. In his account of the logical status of fictional discourse, he claims that 'the author of a work of fiction pretends to perform a series of illocutionary acts, normally of the assertive type' – a class of illocutions which includes 'statements, assertions, descriptions, characterizations, identifications, explanations, and numerous others' (1979: 65, 65n). Despite its lowly status as 'mere' pretence, this act is intentional and self-conscious:

> *Pretend* is an intentional verb: that is, it is one of those verbs which contain the concept of intention built into it. One cannot truly be said to have pretended to do something unless one intended to pretend to do it. (65)

So the writer of fiction is not trying to convince anyone that his words have weight in the world, or carry commitments of the truthful representation variety. The writer may arguably, though, be said to be performing (rather than only pretending to perform) a number of actions, for example: using words, constructing sentences, writing a novel (say) and pretending to assert (which Searle has just claimed is an intentional action). None of these actions are illocutionary acts in themselves, however; rather the use of words and sentences counts as one of the 'lower order or less complex actions which are constitutive parts of [a] higher order or complex action', such as performing an illocutionary act (67–68). Again, the choice of terminology here is instructive: note the sense of a hierarchy of agency which the higher/lower distinction creates. 'Writing sentences' is *part* of what is required – a necessary, constitutive part – for the performance of an illocutionary act, but not *all* that is required. It seems to me that the extra (also requisite) element in Searle's account is intention: the intention to bring into play the 'correct' (i.e., vertical) set of conventions. And, as previously discussed, it is only on the basis of intention that we can discriminate 'pretended' illocutionary acts from the genuine article. In other words, what makes fiction fiction (rather than non-fiction) is not anything intrinsic to the words and sentences (or 'utterance acts') themselves, despite Searle's earlier talk of 'internal' properties, but rather the author's intention to willingly and non-deceptively '[invoke] the horizontal conventions that suspend the normal illocutionary commitments of the utterance' (68).

In an article which followed the first publication of Searle's analysis of fictional discourse in 1974–75, Brown and Steinmann agree with Searle that fiction cannot be defined in 'grammatical (or utterance act) terms' – that is, on the basis of its use of syntax, its phonological or even semantic features – because none of those features are unique to fictional discourse; therefore: 'Whether a discourse is fictional does not [. . .] depend upon what the speaker or writer says; it depends instead [. . .] upon how he intends the hearer or reader to take what he says' (1978: 146, 147–48). So they too discriminate on the grounds of intention ('fictionality has its source in the intentions of the speaker or writer'), and they too are driven to a definition of the fictional 'utterance act' as one that 'pretends' illocutionary force which it does not in fact have: the writer only 'pretends to refer to things' (149, 148). Unlike Searle, however, they introduce the reader's intentions into the equation, arguing that although 'a discourse *is* fictional because its speaker or writer intends it to be so [. . .] [however] [. . .] it is *taken as* fictional only because the hearer or reader *decides* to take it so' (149). Or, we might say, *intends* to read it as such. This decision made by the reader is often made on the basis of textual, but sometimes extra-textual, clues to intention but must be taken even in the absence of such clues. Thus, Brown and Steinmann claim that: 'Lacking evidence of the speaker or writer's intentions [. . .] we must postulate them', otherwise we don't know even whether the discourse is fictional or non-fictional; by which they mean that we don't know which set of constitutive rules are in play (ibid.). In this way they go further than Searle in explicitly arguing the intentionality of all meaning, suggesting that for a piece of discourse to be read as meaningful, it must be read as intentional, even if that intention is somewhat arbitrarily postulated or attributed; we must 'take it' as something (i.e., as either fictional or not, and as meaningful in one way or another). However, the distinction made here between *being fictional* and *being taken as fictional* still suggests that readerly intention to take something as fictional is not enough to 'make' it so.

Because the burden of proof, as far as fictionality is concerned, rests with the author to such a large extent in Searle's account, the fiction/non-fiction and serious/non-serious distinctions which he introduces at the outset of his argument are substantially undermined, particularly given his admission that there is no material difference between these categories, only a difference of intention (which, for Searle, is definitively mental and *immaterial*). In addition, we should note his recognition of the presence of non-fictional elements in fictional works (successfully referring terms such as 'Baker Street' in the Sherlock Holmes novels, and 'England' in many more besides) and the consequent stipulation that in the case of each work of fiction there exists a contract between author and reader concerning 'how far the horizontal conventions of fiction break the vertical connections of serious speech', thereby agreeing what is coherent, allowable or intelligible (for or by the reader) (1979: 73). This implies a continuum – *how far* – rather than the clear and distinct opposition of fiction and non-fiction that he set out to establish. Unlike Austin, this erosion of certainty is not something that Searle is willing to acknowledge.

In brief, Searle is compelled to argue that fictional discourse does not use or comprise 'real' illocutionary acts, because to do so would figure such discourses as essentially indistinguishable from non-fictional discourses. If the difference cannot be located at the syntactical or semantic level then it must be found elsewhere, at the level of the illocutionary. Furthermore, it must be an absence of illocutionary force, rather than just a different use of illocutionary acts, according to Searle. If we were to hold that fiction was comprised of or involved some set of distinct, utterly separate set of illocutionary acts from non-fiction, then this would commit us, unwisely, to the view 'that words do not have their normal meanings in works of fiction' but instead have some new, peculiar set of meanings which would have to be learnt afresh by the reader in each case for the work to be intelligible (64). While it is not immediately apparent that an argument for a distinctive set of illocutionary actions in the field of fictional literature *would* commit us to such a counter-intuitive conclusion, the entailment is brought about by Searle's insistence on a non-contingent connection between meaning and illocution, such that the illocutionary act in any sentence is 'a function of the meaning of the sentence' (ibid.); hence, different illocutionary acts indicate different meanings and hence Searle's confident pronouncement that 'there is no such illocutionary act as "writing a novel"' and the resultant conclusion that fictional discourse does not in fact involve illocutionary force at all (ibid.).

It is on the basis of this – to my mind rather unstable – argument that Searle presents us with the alternative view of fictional texts as 'pretended' illocutionary acts, thus reinforcing the ordinary language/literary language distinction and endowing (fictional) literature with a highly problematic 'otherness'; the possible continuity or contiguity of fiction and non-fiction is something he will not countenance. Neither lying nor truth-telling, fiction instead inhabits a kind of ontological no man's land, being a matter of intentional but 'nondeceptive pseudoperformance' (65). Fictionality, it seems, is structured around absence: an absence of force with a consequent abnegation of 'real-world' responsibilities. How can such terminology not be read as instituting or at least reinforcing a negative hierarchy of the literary and the ordinary? Even those philosophers who dispute Searle's account of fictional discourse frequently do so in a manner which actually consolidates this opposition – even if questioning the hierarchical ordering of this. Kendall Walton is a case in point: arguing both that assertions may be contained within fictions and that pretence alone is insufficient to determine fictionality, his primary objection to any application of speech act theory to literature is nevertheless that it is missing the point to understand fictional uses of language 'in terms of their use in making assertions, asking questions, issuing orders, and/or engaging in other activities characteristic of nonfictional language', thus suggesting that fiction is engaged in some completely other kind of business than communication – the business of providing aesthetic pleasure or satisfaction, perhaps (1983: 84). The admission that this is, crucially, a lesser kind of business than that of communication (had we not already imbibed that familiar message, on the basis of the very different social, cultural and political resonances of the very terms 'aesthetic' and 'communication') is consolidated in Walton's remark that fiction '*is* perhaps parasitic on "serious" discourse' (ibid.). Interestingly, like Searle, Walton postulates a 'disanalogy between illocutionary actions and fiction-making' on the grounds of the different intentions governing them, but does so through the claim that the agent's intention is more crucial to the understanding of and response to an illocutionary action than for the comprehension of a fiction (86). This is part of his more general claim that 'the actions of artists' should not be placed 'at the centre of one's theory of fiction' (87). So he is an anti-intentionalist about literature and an intentionalist about 'ordinary' speech, the adoption of which position further polarizes those problematic categories, as well as implying that the 'application' of speech act theory to literature, even if this is possible, will not necessarily answer the demands of the intentionalists. If Walton's logic is accepted, then to argue for literary intentionalism on the basis of the intentionalism of ordinary (non-literary) speech is to make a peculiar sort of category error.

Gregory Currie, too, despite offering a reasoned opposition to Searle's mimetic theory, falls into a similar trap regarding the definition and relative status of the poetic and the ordinary. For Currie, the writer of fiction *does* actually perform – rather than merely pretending to perform – an illocutionary act, but it is a 'fictive illocutionary act' (1986: 305); whether or not this lack of pretence makes it 'genuine', he does not say, but he does show some polemical verve in his stated desire to 'treat the utterance of fiction as the performance of an illocutionary act on a par with assertion' (1985: 385). Notwithstanding this postulation of a fictional speech act that is 'distinctively different from' and yet seemingly as valid as the non-fictional acts of speech to which it bears some structural similarity (one in the eye for the theorists of the 'parasitical'), Currie yet strays dangerously close to Beardsley's account of fiction and illocution in his claim that 'the utterer of fiction [. . .] wants to get the audience to *make-believe* the proposition uttered' (1986: 304). How does the make-believe theory differ from the mimetic, in real terms? Well, although the reader is encouraged to 'make-believe' the proposition, the utterer is not himself engaging in an act of make-believe but rather in a 'real' act (which is the act of encouraging the audience to *make-believe* the proposition, as contrasted with the act of encouraging them to *believe* it, as a writer of non-fiction would be likely to do). So Currie also makes the fictional/non-fictional distinction on the grounds of intention: rather than intending to suspend illocutionary force, as Searle's writer of fiction elects to do, Currie's hypothetical poet intends to produce a fictive illocutionary act, rather than some other kind, and fictional texts are, therefore, 'the products of an utterer with certain kinds of intentions and beliefs'; namely, the intention to get the reader to make-believe some proposition or other by means of a fictive illocutionary act (ibid.). Furthermore, the reader, in order to participate fully in the 'make-believe', must recognize these intentions – and their awareness of the 'make-believe' status of the assertions made suggests that it is not the author but the reader who is engaging in 'pretence'; the author's actions are transparency itself. Indeed, 'what makes [a fiction] fiction is the [speaker's] intention to get the audience to make-believe via the Gricean mechanism', that is: to get the audience to recognize his intention that they make-believe (305).

What is notable here is a particular insistence on the recognition of authorial or speaker (rather than any other) intention to ensure the success of the make-believe (and, indeed, to ensure its very make-believe*ness* – if the intention is *not* correctly identified then the audience is simply misled as to the status of the illocutionary acts that they are interpreting and therefore as to what kind of communicative practice they are engaged in). This is revealed in Currie's claim that 'fictional works cannot be treated as disembodied texts', but rather 'must be identified as the products of an utterer with certain kinds of intentions and beliefs' and leads to much troublesome talk of (presumably 'private' and only incidentally accessible) beliefs, conditions of sincerity and felicity, as evidenced by the following protestations regarding 'truth' in fiction:

> What determines what is true in fiction is a certain pattern of inferences that take place within the scope of the reader's make-believe. We make-believe that the story is told to us as assertion by someone who shares the common beliefs of the society in which the work is written. We then use the text and the background of common beliefs to work out what this person believes. What it is reasonable to infer that he believes is exactly what is true in the fiction. (307)

'Truth' in fiction, then, is something which must be placed there by some 'teller', hypothetical or otherwise, presumably to set parameters for our reading of the text. But it is not always evident who, if anyone, is 'telling' us the story (not all poems and novels adhere to a clear scheme of first- or third-person narration), and Currie does not argue the case for the necessity of an assertorial character for fiction. Why do we need to infer the 'beliefs' (for which read 'intentions') of the teller when we have the text itself as evidence for its 'own' truth? Why, in any case, do we need to think of it in terms of 'truth', when doing so seems always to underline the purportedly parasitical nature of fictional discourse? Even if the teller is identified as being 'internal' to the text, then it is not clear why we need to hypothesize such a person or function in the first place. In addition, the woeful under-determination of such key factors in Currie's equation as 'the common beliefs of the society' and 'what it is reasonable to infer' is obvious enough to not be worth going into here. And again, Currie does not address the problem of how we tell fictive from non-fictive illocutions in cases where the utterer's intentions are not clearly signalled. Despite the unwieldy extra baggage of fictional 'tellers' and 'fictional truth', the make-believe theory is only a step away from the mimesis theory of literary speech acts: an illocution *is* performed, *but* it is a fictive, non-standard one and, to that extent, barely an illocution at all; the barrier between poetic and non-poetic discourse remains impermeable.

Given the prevalence of these arguments for the otherness of the aesthetic (or categories which are seen as supervenient upon it, such as the literary) it is no wonder that Searle's fellow analytic philosophers (and sundry aestheticians) should be led to query exactly how such an ethereal entity as fiction (either an action, or the product of an action which is without force and without commitments or purchase in the world) should stimulate any kind of real world emotional or empathetic response in us.[7] Do we take pleasure only in the skill of the performance or the uncanniness of the pretence (its *realism* perhaps, rather than its *reality*), thus reducing fiction to pure form or – in the terminology of this argument – bare locution?

Literature as 'speech context': Towards a SAT of literature?

Stanley Fish makes the point, in his response to Austin and Searle, that 'the distinction between serious and fictional discourse [. . .] cannot be maintained if the implications of speech-act theory are clearly and steadily seen' (1980: 197); the deconstruction of the poetic/ordinary language opposition is therefore the requisite first step in Mary-Louise Pratt's journey *Toward a Speech Act Theory of Literary Discourse* (1977). In fact she demonstrates the erroneousness of such distinctions as literary/ordinary, serious/non-serious and so on, as a means of arguing the relevance of speech act theory – and other theories of everyday linguistic usage – to the study of literary texts. As part of her argument that literary utterances exist on some kind of linguistic continuum with conversational and other communicative utterances, she cites in particular: the 'literariness' of so-called everyday linguistic exchanges; the presence of speaker and audience even in literary speech situations; and the knowledge of shared rules and conventions that is a feature of both literary and non-literary speech situations. The suggestion of this linguistic continuum on which the position of literary utterances can be plotted leads her to dismiss those formalist approaches to literary texts which have presented a view of literature as 'linguistically autonomous' and as 'formally and functionally distinct from other kinds of utterances' (1977: xii). Pratt reads the opposition as being one between 'formal' and 'sociological' approaches to the text; the emphasis of speech act theory upon the social contextualization of speech, and the socially situated uses of speech, illustrates on which side Pratt's allegiances are likely to fall, despite her stated intention of 'sealing the breach' between the two sides (xix).

This depiction of language as 'social process' rather than 'formal structure' is also what appeals to a more recent advocate of a speech act theory of literature, Sandy Petrey, whose desire to emphasize the conventional (i.e., collectively determined) nature, function and meanings of literature expresses an ideological imperative to '[bring] together the inner self and the outer world, the individual and the communal', thereby identifying literature as one of the (many) ways in which we 'participate *perceptibly* in communal life', as a facilitator of that 'communal life', rather than some idiosyncratic component of it (1990: 3, 6).

In the course of noting the 'literariness' implicit in 'ordinary' language – evident, for example, in the use of narratives and of figurative language as part of our day-to-day communicative practice – Pratt contends that:

> All the problems of coherence, chronology, causality, foregrounding, plausibility, selection of detail, tense, point of view, and emotional intensity exist for the natural narrator just as they do for the novelist, and they are confronted and solved (with greater or lesser success) by speakers of the language every day. (1977: 66–67)

Even bus stop conversations have 'themes' and employ 'narrative techniques', although we may not always recognize them as such. She therefore concludes that features customarily identified as 'literary' by critics 'occur in novels not because they are novels (i.e., literature) but because they are members of some other more general category of speech acts', so that the 'poetical' and our responses to it – the 'hypothesizing' or active interpretation that we may be required to engage in – can be seen simply as characteristic of language use in general, evident in many different situations in which language is used (69). Literature, according to this account, is merely one among many different uses of language or can be seen as constituting one of many different situations or contexts in which language is used.

The erosion of the poetic/ordinary distinction stands as Pratt's primary justification for the application of speech act theory to literary texts, and she is subsequently able to cite the relevance to literature of that theory's emphasis on context, the intentions of speech situation participants, the relationship between those participants (whether figured as speaker/hearer, or as author/ reader) and the rules which structure – and, indeed, enable – that relationship:

> Literature itself is a speech context. And as with any utterance, the way people produce and understand literary works depends enormously on unspoken, culturally-shared knowledge of the rules, conventions and expectations that are in play when language is used in that context. (86)

While Searle's insistence on the rule-governed nature of all 'genuine' speech acts provided the basis for his exclusion of fictional discourse from this category, Pratt brings to light the centrality of convention(s) to literature. Petrey argues the folly of Austin's denial of illocutionary force to literature from a similar standpoint: 'The entire thrust of Austin's thought is that words perform by virtue of the conventions social being applies to them. That fundamental insight is not compatible with his opinion of literature' (1990: 52). Yet literature is, affirms Petrey, 'a thoroughly conventionalized [. . .] language form' (ibid.). Certainly this coheres with our idea of Literature (with a capital L) as a *practice* which is, to a large degree, collective or communal, or a *category* which is both collectively determined and which serves some broader social purpose – rather than a *thing*. In literature, the kind of contextual and conventional information required for successful 'communication' could range from the most basic recognition that the text under consideration is fiction, to a more sophisticated knowledge of genres and subgenres, of relevant historical periods (knowledge of literary history and the chain of influences), and the various uses of certain figurative tropes. In fact, Elizabeth Traugott has given an account of literary genres as 'systems of appropriateness conditions', which stipulate what is permissible (in both form and content) and therefore what qualifies a work for membership of a particular genre (qtd. in Pratt 1977: 86); these conditions can be seen as analogous to the kinds of conditions for successful speech acts set down by Austin and Searle. While rules of literary genre may be argued to form a distinct, specialized set of rules (although Pratt actively disputes that here), there is no reason for them to be 'horizontal', as Searle claims; or if they are, then the 'genuine' speech acts that he considers are governed by rules which are no less 'horizontal'. It is not only fictional utterances that suffer an anxiety of reference.

So, despite being warned off explicitly by Austin and Searle, and implicitly by Grice's preponderant concern with propositional content, both Pratt and Petrey argue the case for a speech act theory of literature. In fact, Pratt works to adjust Grice's Cooperative Principle to fit it to literary utterances, turning her attention in particular to the rule of Relation which urges speaker relevance. For relevance, Pratt substitutes the concept of 'tellability', claiming that an utterance is 'tellable' if it asserts or conveys something interesting or notable or 'problematic'; this, she suggests, constitutes a kind of 'display-producing relevance', rather than 'the kind of relevance we expect of assertions made in answer to or in anticipation of a question, these [latter] being examples of what Grice means by a "maximally effective exchange of information"' (1977: 136). She further clarifies 'tellability' by explaining that 'Assertions whose relevance is tellability must represent states of affairs that are held to be unusual, contrary to expectations or otherwise problematic; informing assertions may do so, but they do not have to, and it is not their point to do so' (ibid.). According to this logic, literary works are examples of this 'verbal experience-displaying, experience-sharing activity', and are intended as 'display texts' rather than 'informing texts' (140, 145). It is intention rather than semantic or syntactic elements which makes the difference, as 'tellability' is a characteristic that may be found in display texts and informing texts alike: as Pratt claims, it is the 'point' or purpose of display texts to be subject to the laws of 'tellability'.

The flaws in this theory are fairly obvious, not least in the way it appears to reassert a dichotomy of the literary and the non-literary, implicit in the distinction between utterances (or texts) which are intended to 'display' and those which are intended to 'inform' (even though the latter distinction will not be identical to the former). Is the 'point' of literary utterances only to 'display'? Pratt suggests that this is their pre-eminent purpose. In addition it leads to her counter-intuitive proposal that literary works are always about people in situations that are 'unusual' or 'problematic' or that such works always explore 'emotive' experiences or states and – moreover – that this is what makes them interesting and culturally significant (140–41). While we may not lament this preclusion of any kind of literature of the banal, yet the erroneousness of this linking of the 'emotive' and the culturally valuable and valued cannot be ignored. It is symptomatic of a sociological approach to literature, which ultimately values it in purely sociological terms: as a social document (in particular here, as *expression*), designed to teach us about ourselves and our interaction with others in society; this is to the detriment of any consideration of those pleasures in reading which tend more towards the aesthetic, or the sensuous, than the pedagogical or informative.

There are other, more general, problems in Pratt's account, which perhaps point up inevitable obstacles to the application of speech act theory to literature, and so to the manipulation of one kind of intentionalism into another. In asking whether a literary text comprises a series of speech acts or a single one, Pratt introduces the notion of the 'multisentence utterance' as something which might be applicable to texts (85). This is useful up to a point: most speech act theorists address themselves to single sentence utterances, and most attempts to bring together speech act theory and literature look only at speech acts *in* literature (either the direct speech of characters in a novel, or the pronouncements of its narrator, or the dramatic dialogue in a play) rather than looking at a literary text as a whole *as* a speech act or series of speech acts. Pratt's notion of 'multisentence utterances' allows us to do this, but is immediately problematized by her claim that such utterances will have 'appropriateness conditions that [. . .] apply across the entire utterance', in virtue of the utterance's 'single point or purpose' (ibid.). Aside from being an obviously reductive characterization of a literary text, this heralds a return to monism, with its talk of *the* (single) meaning or message of the work, and is therefore to be avoided at all costs.

There is also a degree of essentialism in Pratt's approach which means that, despite her avowed antagonism to the poetic/non-poetic language distinction, she still writes about the 'essence of literariness or poeticality' which is seen to reside 'not in the message but in a particular disposition of speaker and audience with regard to the message, one that is characteristic of the literary speech situation' (87). In this way a distinction is still made between the literary and the non-literary 'speech situation' which posits literariness as non-standard *usage* rather than non-standard *type* of language. If we concentrate on the similarities between and overlap of literary and non-literary speech, as she urges us to do, and if we further recognize – as Pratt fails to do – the difficulties of exhaustively determining context and 'the communicative purposes of the participants' (the two factors which she claims do allow us to differentiate between fiction and non-fiction) it surely becomes more and more difficult to use concepts such as 'literariness' (88).

The major problem with Pratt's theory may be the way in which the literary, ultimately, is figured as a subset of the linguistic, which is suggestive both of its second-order status, and of its definability as a subset (the very notion of 'set' implies that something either will or will not belong to it). In fact it is to institutional factors that Pratt appeals in her normative definition of literature as *that which gets published* – because it is the 'best' and 'most interesting' writing – witness her claim that 'literary selection conspires always to eliminate what a community takes to be the least interesting and least skilful verbal productions' (124). She does not address the question of who makes the selection, or delineate the criteria according to which a work is 'good' or 'interesting'. This problem is, I suppose, also a feature of pragmatic accounts of literature, such as that of Stanley Fish, which suggest that 'literature' is just those texts that we, as an interpretative community, choose to call by that name or classify under that heading; if 'literature' is still seen as an evaluative term, then the self-congratulatory nature of this definition – even where it concedes the relativity to a particular interpretative community in each case – might give us pause for thought.

One further flaw in Pratt's account is her suggestion of the non-participation of audiences – and by analogy, readers – in the literary speech situation. She writes of 'the particularly silent and solitary nonparticipation' which such a situation imposes, comparing the reader to a muted, passive, captive – and consequently judgemental – audience at a public lecture or in court (111). This elides the role of the reader in actively construing meaning and instead figures him as willingly giving up the floor to the speaker whose authority is – temporarily at least – assured; the reader's 'judgement' is silent, or at least indefinitely postponed.

Sandy Petrey is highly critical of Pratt for what he sees as her resurrection of the author in order to '[collapse textual illocution] into face-to-face dialogue' (1990: 77). He prefers to employ speech act theory as the pretext for a theory of texts as socially situated and hence for interpretation as a communal (conventional) type of activity, something altogether more democratic than the didactic court room or lecture hall scene that Pratt has in mind. In Petrey's opinion, 'literature, like any other linguistic performance, is a collective interaction as well as a verbal object', and therefore speech act theory is not only applicable to literary texts, but – unusually for a critical method – provides an opportunity to practice a kind of 'socialized criticism', by considering 'the societies from which [literature] comes and in which it circulates' and by relating texts to 'social reality' (70). For Petrey, this constitutes an ethical approach to the text analogous to that of E. D. Hirsch, although this is not a debt that he fully acknowledges; neither does he justify the assumption that giving pre-eminence to a conceptualization of the literary text as 'collective interaction' is somehow more ethical than considering it as a 'verbal object'; similarly, he attacks structuralism and transformational grammar alike for their focus on 'language in itself' rather than 'language in use' without justifying the suggestion that attention to use or interaction is inherently more 'ethical' than attention to form (48). This further suggests that 'language in itself' is being conceived here as pre-intentional and, in some fundamental way, inanimate or inhuman. But while Pratt and Petrey both figure force as the vital human addition to bare language or locution, the latter argues more specifically that illocutionary force 'derives from *collective* protocols' – that is, it is the result of collective rather than individual agency – and this serves as his refutation of Pratt's emphasis on the author: 'Every speech act is an institutional fact; none requires the brute fact of a particular person's presence in a particular place' (79).

This is also his way of establishing the utility of speech act theory for cases of *written* discourse where the author is likely to be absent, as he further claims that 'the speaker's presence is an illocutionary irrelevancy' (79–80). Such a declaration would seem to preclude the problems raised by the blueprint conception of intention, as the context of reception is privileged over the context of creation. Unfortunately, though, Petrey's banishment of the individual author/speaker/agent figure, while making the theory more applicable to literature, takes it some distance from Austin's initial definitions; one of the speech acts that Austin takes as a paradigmatic performative – saying 'I do' at a wedding ceremony – does require that certain people be present for illocutionary uptake to occur. On the basis of the possible absence of the (literary) speaker, Petrey goes on to argue the concomitant irrelevance of the speaker's intentions, but the intention that he is dismissing is the 'inward or spiritual act' of Austin's musings on the subject (1962: 9), and he rather reductively characterizes the 'intentionalist vision of literary analysis' as advocating the discovery and public delineation of this definitively private act (1990: 80). This is an assessment of their aims which most contemporary intentionalists would immediately reject, as documented in the closing sections of the last chapter.

So, then, while Petrey's banishment of the individual, omnipotent author is an improvement on Pratt's agent-centred application of speech act theory to literary utterances, still he will not countenance the existence of any kind of collective intention, pertaining to his Fish-inspired critical communities. Nor will either theorist entertain the idea of an intentionality of form – indeed Pratt at every stage resists and refutes the idea that the text might be intent upon its own meaning. This she dismisses as an error of Russian Formalists such as Roman Jakobson:

> The well-known term 'dominant' was frequently used to convert author intention into an aspect of the text [. . .] the dominant is the 'focusing component' which 'specifies the work' [see Jakobson 1987: 41]. The overriding tendency to disguise all notions of intention, perception, and value by converting them into textual attributes has a conspicuous stylistic effect on almost all formalist and structuralist writings. (1977: 74)

Not only this, but such *incorporation* of intention (literally: making it part of the *body* of the text) leads, says Pratt, to an 'impoverished' view of poetic language because it makes 'literary composition' seem like 'a kind of self-motivated automatic writing dependent only incidentally on the participation of human beings', and for this avowedly humanistic reason, she rejects outright a view of the text which she sees as 'mechanistic' in the way it '"forms itself" and "orients itself" according to its own intentions and values' (75). This emphasis on the human and sociological aspects of language use, which figures that language use as expressive of our humanity and inseparable from it – remember the distinction made between 'human *action*' and 'mere animal behaviour' by Max Black (1973: 258), to which we might add as equally 'undesirable': mere automatism – this is ultimately what limits the usefulness of speech act theory for my purposes.

Conclusion

Intention, for the speech act theorist, will always be something outside the text, locatable only by means of an extrinsic criticism which considers the situations in which speech occurs and the ways in which speech is used rather than speech in itself, or in abstraction. In fact, speech in abstraction is an utterly alien notion to a speech act theorist, deliberately excluded from the discussion because of the mechanizing or dehumanizing of the text that it is seen to imply. Speech act theory must, then, be read as having some kind of ethical imperative, in its emphasis on the social and collective contexts of speech, and in most cases (Petrey being the possible exception) when applied to literary texts, as having an undeniable investment in the authorial delimitation of meaning. 'Clearly', Pratt assures us:

> it is the reader who focuses on the message in a literary speech situation, not the message that focuses on itself. Likewise it is the speaker, not the text, who invites and attempts to control or manipulate this focusing according to his own, not the text's intention. (1977: 88)

But the 'clarity' of this account of literary meaning, where the author is invested with the requisite intentional agency for which the text is merely a vehicle, is precisely what I would like to throw open to doubt.

Ultimately an account of the literary text as speech act or utterance has been revealed as an over-determination of what literary texts are and how they mean: primarily in its emphasis on (illocutionary) *force*, to the exclusion of (locutionary) *form* and, significantly, in its tendency to identify the author as the sole originator of force (and, therefore, meaning). Despite Hirsch's optimistic appropriation of the language of speech act theory in 1975, it has become evident that an identification of illocutionary force and intention does not help us to move away from a conceptualization of intention as either prior blueprint or *a posteriori* addition, always troublingly dissociable from the form of the text or utterance itself. Nevertheless, speech act theory is independently useful in its foregrounding of intention and convention – and the delicate relationship of these two – as primary players in the production and understanding of meaning, and in positing literary utterances as socially situated; even if the more radical ramifications of this are resisted in the adherence to a rather undeveloped notion of intention as, necessarily, something resident in the mind of the author.

In the final chapter the deconstruction of the binary figuration of form and force will be realized as the necessary first premise for the theorization of an intentionality of form which might posit a 'third way', between or beyond a consideration of literature as either primarily formal or primarily social. In her rejection of the lessons of Russian Formalism, Mary-Louise Pratt envisages the gulf between such formal and sociological approaches to literature as fundamentally unbridgeable, but the fact that this rejection is based upon a reductive choice between the 'human' and the 'mechanical' might lead us to hazard a little optimism here. In the meantime, some useful re-thinking of the notion of intention is available to us through a reading of phenomenological theories of intentionality: theories which may offer some rejoinder to the persistent complaint of intention's inaccessibility by clarifying the relationship between mental act and mental object when the object in question is a literary text.

Notes

1 N. B. the distinction as stated assumes that form is, by definition, pre-intentional or at least independent of intention; this is highly debatable. Ultimately I want to consider a kind of intentionality of form which will incorporate within its remit some notion of force and so demonstrate that form and force are inextricable, but that will be the subject of the final chapter.

2 Not only the earlier mentioned literariness but also the fact that the posthumous published version is taken from Austin's own notes, made over the course of a number of years, and designated for a number of different lectures at Oxford and Harvard; the notes are in places 'fragmentary' with 'very abbreviated' marginal additions; the finished text was compiled and edited by J. O. Urmson and Marina Sbisa, the former conceding that 'these lectures as printed do not exactly reproduce Austin's written notes' (Urmson 'Preface to the First Edition', in Austin 1962: vii).

3 Grice distinguishes between 'natural' and 'non-natural' meaning. The former covers cases such as 'those spots mean measles' (Grice's example), where 'x means that p' states a relationship of entailment between x (spots) and p (measles). The latter covers cases such as 'those three rings on the bell mean that the bus is full' (also Grice's example), where the relationship between x and p is conventional, agreed upon, but not one of entailment (1989a: 213–15). The bus bell example reveals another point of interest in Grice's work: his use of 'utterance' as 'a neutral word to apply to any candidate for [nonnatural] meaning', such that it denotes not only the category of the linguistic, and not only what can be described as 'acts', but rather 'has a convenient act-object ambiguity' (216). This expansion of 'utterance' makes it, I think, more ripe for appropriation by literary theorists.

4 Note the implied interchangeability of these two terms, which gestures towards a notion of intention as intentionality, in the phenomenological sense of that term. The next chapter will address phenomenological theories of intentionality in detail.

5 N. B. Kendall Walton (1983) distinguishes between this and the 'pretence' account which he attributes to Searle, but I would contend that there is sufficient overlap between the two approaches to merit classing the latter as a form of mimesis, although Searle himself dismisses the term as unhelpful: 'It is easy to stop thinking about the logical status of fictional discourse if we repeat slogans like "the suspension of disbelief" or expressions like "mimesis". Such notions contain our problem but not its solution' (1979: 60–61). The fact that notions such as 'mimesis' 'contain' the problem, I am taking as sufficient evidence for the use of the term to classify his approach; also, the suggestion of the lack of illocutionary force attendant upon fictional utterances (e.g., the non-seriousness of such utterances and their lack of real world commitments) is reminiscent of the original Platonic 'degrees of reality' theory in which the idea of artistic mimesis has its inception.

6 Which is to say: non-fictional speech acts, either written or spoken. The example Searle gives of a 'genuine' speech act, or series of speech acts, is a passage from the *New York Times* and so he cannot, on this occasion, stand accused of phonocentrism.

[7] See, e.g., O. Hanfling, 'Fact, Fiction & Feeling', *British Journal of Aesthetics*, 36.4 (1996): 356–66; K. Walton, *Mimesis of Make-Believe* (Cambridge, MA.: Harvard University Press, 1993); K. Walton, 'Fearing Fictions', *Journal of Philosophy*, 75 (1978): 5–27; P. Lamarque, *Fictional Points of View* (Ithaca: Cornell University Press, 1996); C. Radford, 'How Can We be Moved by the Fate of Anna Karenina?', in A. Neill and A. Ridley (eds), *Arguing About Art* (London: Routledge, 2002), pp. 239–49; A. Savile, 'Imagination & the Content of Fiction', *British Journal of Aesthetics*, 38.2 (1998): 136–49; R. Joyce, 'Rational Fear of Monsters', *British Journal of Aesthetics*, 40.2 (2000): 209–24; and J. Harold, 'Empathy with Fictions', *British Journal of Aesthetics*, 40.3 (2000): 340–55.

Chapter 3

Intentionality: Meaning and the Mental

Introduction: Hirsch and phenomenology

If we hark back to Wimsatt and Beardsley's original banishment of authorial intention in 'The Intentional Fallacy', we can recall that their attack was two-pronged, asserting as it did that what was going on in the mind of the author at the time of writing is not *desirable* information for the critic but also, crucially, that it isn't *available* information (1962: 92). I have suggested throughout that this identification of intention and mental content is problematic – in fact Wimsatt and Beardsley themselves offer an alternative reading of intention as something that is 'effective' in the text. Nevertheless, the debates around intention have been dogged by more or less philosophical questions about the accessibility (or not) of 'other minds' – the authorial mind in particular – and by questions about the extent to which any discussion of literary meaning ultimately reduces to a discussion about consciousness and cognitivity. Is meaning in the text or in the mind? If the latter, whose mind is pre-eminent here, the author's or reader's? More importantly, can these issues be considered without recourse to this uncomfortable disjunction of mind and text: might texts themselves be possessed of something describable as a 'mentality' and, therefore, of intentions in their own right? In the chapter that follows, I'll ponder these questions via interrogative readings of, first, Hirsch's engagement with phenomenology and, second, phenomenological theories of intentionality and reading; the latter may help us towards those twin goals of reconceptualizing intention and resituating it within the text.

In *Validity in Interpretation*, Hirsch defends himself abrasively against all psychologistic objections to his intentionalism, declaring that: 'It betrays a totally inadequate conception of verbal meaning to equate it with what the author "has in mind"' (1967: 18). His key contention that 'meaning is an affair of consciousness not of words' (which, despite his elaborate type theory of meaning, appears to privilege intention over convention) rests on a distinction between mental *processes* and mental *objects* (4). He writes of consciousness that 'the objects of its awareness are not the same as the subjective "perceptions", "processes", or "acts" which are directed toward those objects', and in the third Appendix of the book he describes a type as being a 'mental object' or 'idea' (37, 265). This hints at the psychological aspect of meaning reconstruction for Hirsch, the extent to which it is consciousness that is constitutive of meaning. Indeed there would be *no* meaning without consciousness: 'there is no magic land of meanings outside human consciousness', he states (4). Meaning is not, therefore, identified with mental processes by Hirsch, but neither can it exist apart from them.

To elucidate this act/object distinction, Hirsch borrows from Husserl, and so brings into play a different conception of intention as the relation between mental act (what Husserl would eventually term 'noesis') and the object ('noema') towards which it is directed and intentionality as this 'directedness' of mental states upon certain objects. As William Ray affirms, 'phenomenological theories of reading [. . .] enclose both act and structure within a single concept: that of *intention*' (1984: 8). Intentionality in the phenomenological sense stipulates, first, that consciousness is always consciousness *of* something and, second, that consciousness is an *act* by which a *subject* intends an *object*, and 'intention' here is synonymous with 'means, imagines, conceptualizes, is conscious of' (ibid.). This phenomenological approach may serve to counter objections concerning the alleged inaccessibility of authorial intention, because it allows for different mental states (or the mental states of different individuals) to 'intend' the same object and in that act of intending it stresses the connectivity of act and object, subject and world. David Michael Levin's explanation of the act–object distinction within phenomenology, in his introduction to Roman Ingarden's *The Literary Work of Art*, perhaps helps to reinforce the uses of phenomenology for literary criticism. His claim that 'it is of the essence of an object that it is logically distinct from (i.e., logically irreducible to) the mental act that is occupied with it' confirms our intuitions that the text, even when viewed as an 'intentional object', is logically distinct from the consciousness of both author and reader (and thus not wholly imaginary or immaterial), and that it has some kind of existence beyond their respective apprehensions of it (1973b: xix). His assertion that the text 'can become *the self-same* object of an infinite number of mental acts' and that these might be different kinds of acts (perceptual, desiring, etc.), and thus that the object 'is *transcendent* with respect to the multiplicity of its logically positive correlative intentional acts', and so 'is never accessible in its absolute totality of properties' is also useful for our purposes (ibid.). Again it confirms our intuitions that multiple possible readings of the same text will not necessarily trouble the sameness/self-identity of that text, and that any perception we might have of that text is necessarily incomplete – we cannot access it in its totality. Phenomenological approaches to literature thus go some way towards confirming and accounting for the particular ontology of the literary text and the jointly objective and subjective nature of literary criticism.

However, it is intentionality which particularly concerns me here, and it is Franz Brentano who first develops the theory of intentionality, an idea borrowed from Scholastic philosophy. Although he doesn't use the term 'intentionality', he does write of intentional objects and the intentional relation between mental act and object:

> The common feature of everything psychological, often referred to, unfortunately, by the misleading term 'consciousness', consists in a relation that we bear to an object. The relation has been called intentional; it is a relation to something which may not be actual but which is presented as an object. There is no hearing unless something is heard, no believing unless something is believed; there is no hoping unless something is hoped for [. . .] and so on, for all the other psychological phenomena. (1969: 14)

This notion of intentionality as the directedness of all mental states and acts is subsequently taken up and adapted by Husserl and it is the Husserlian version that feeds into various more or less 'phenomenological' theories of reading. It is important, nevertheless, to distinguish *intentionality*, in the phenomenological sense, from *intentionalism* in the literary critical sense; these terms are certainly not interchangeable. The former does not offer a straightforward explication of causal or creative origins, neither does it figure intention as a kind of will to act or mean, but rather, as stated earlier, as a *relation* between acts and objects. It does, however, posit the necessary existence of an intending subject, *for whom* and as a projection of whose consciousness the object exists; it also argues the fundamentality of the mental, as all experience is first and foremost mental experience; everything else can be 'bracketed' off, according to the practice of phenomenological 'reduction'. In Husserl this process of paring down is fundamental to the practice of phenomenology, hence his insistence upon the '*principle of presuppositionlessness*', the claim to concentrate on 'careful description of phenomena themselves, to be attentive only to what is *given* in intuition' – so attention is paid to things *as they appear* rather than *as they are in themselves* (indeed the latter, insofar as it *can* be known, can *only* be known by means of the former, the world of appearances) (Moran 2000: 9). All that can be described and known is what appears to consciousness, in the way that it appears to consciousness: in other words 'all experience is experience *to someone*', some conscious subject (Moran 2000: 11). And consciousness knows itself with an immediacy, a self-presence, which guarantees that knowledge and makes it, in its indubitability, the foundation of all other knowledge that might be possible.

So, in the place of intention as one mental state among many possible mental states, we have intentionality as a characteristic of *all* mental states, because all of these states display a kind of 'aboutness', which is to say that all are directed upon some object: fear is fear *of* some thing, desire is desire *for* some thing and so on. We can note in Ray's definition of this theory of intentionality the mutual dependence of subject and object: the intentional object, that which is perceived, imagined, intended, is 'contingent on the subject for its givenness' (1984: 8). It is this which has led to allegations of idealism and/or solipsism against phenomenology, however 'the intuition of that [intentional] object constitutes the subject as an awareness' – that is, as that which intends (means, imagines, conceptualizes, etc.) (ibid.). The subject, then, is a subject only insofar as it intends some object, and these objects 'exist' only in the intentional purview of some intending subject, so the relationship is reciprocal. However, it should be noted here that by 'exist' I mean only 'come into givenness'; the objects in question may be *either* real or imaginary, it is their 'givenness' to perception, to the mind, which counts, not their location in some kind of extramental 'real' world; it is a matter of perception or consciousness rather than ontology. It is at this point that the allegations of idealism frequently wielded against phenomenology *may* surface, if the 'real' world is seen as supervenient upon the phenomenal world, rather than vice versa.

Dermot Moran offers a response to such allegations in his explanation of the reciprocal nature of the subject/object relation, as follows:

> Central to phenomenology [. . .] is its attempt to provide a rigorous defence of the fundamental and inextricable role of subjectivity and consciousness in all knowledge and in descriptions of the world [. . .]. But [. . .] [this] is not a wallowing in the subjective domain purely for its own sake. Indeed, the whole point of phenomenology is that we cannot split off the subjective domain from the domain of the natural world as scientific naturalism has done. Subjectivity must be understood as inextricably involved in the process of constituting objectivity. (2000: 15)

Objectivity is thus always 'objectivity-for-subjectivity' (ibid.); this is Husserl's formulation in the *Logical Investigations.* Is it then a 'branch' or variant of subjectivity? Can the distinction between subject and object – so crucial to phenomenology – in fact be maintained if the latter exists only 'for' the former? Moran, for one, is not prepared to countenance the disintegration of this distinction, continually emphasizing 'the worldliness of consciousness' for Husserl, and the way in which phenomenology attempts to plot a middle course between rationalist and idealist versions of reality; the accusations of subjective idealism are unfounded, according to this reading (12). Nevertheless, Sean Burke warns against misreadings of Husserlian phenomenology in the other direction when he takes issue with Foucault's description of it in the following terms: 'The phenomenological project continually resolves itself, before our eyes, into a description – empirical despite itself – of actual experience' (Foucault 1970: 326). Burke is highly sceptical of such a characterization:

> The inference we are to make here is, presumably, that since the intentionality of consciousness, as understood by Husserl, must be consciousness of

> something, then phenomenology was bound to predicate an extramental, empirical realm. But the predication of such a realm is by no means tantamount to its empirical description, and to call a system 'empirical' which (however unsuccessfully) brackets off that realm in the interests of elaborating a pure philosophy of consciousness, involves a considerable extension of what we understand by an empirical science. By the same criteria, any system which incorporates some acceptance of a real, physical world exterior to consciousness would be empirical, or nearly so. (1998: 73)

This matter – of phenomenology's fraught and disputed relation with the 'real world', and of the mind's relationship with its objects – will be brought into starker relief, and greater relevance, if we consider it when the intentional 'object' in question is a work of literature or the set of possible meanings of which that work is comprised.

The uses of the phenomenological approach for Hirsch may, then, become evident. Textual meaning is now figured as an intentional object – something towards which consciousness may be oriented – and an unlimited number of intentional acts may intend the same meaning. In other words, the interpreter may, by a process of psychological reconstruction, intend the same object (meaning) as the author, and so correctly interpret the work. Thus, according to Hirsch, 'the interpreter's primary task is to reproduce in himself the author's "logic", his attitudes, his cultural givens, in short, his world', and 'the ultimate verificative principle is very simple – the imaginative reconstruction of the speaking subject', or, at least, 'a very limited and special aspect of the author's total subjectivity', namely, 'that "part" of the author which specifies or determines verbal meaning' (1967: 242–43). This is what it is to *re-cognize* the author's meaning: to direct your mind in the same way and upon the same objects, to emulate authorial cognition. Hirsch further reinforces this sharability – that is, the possibility of such reconstruction or re-cognition – by his emphasis on the self-identical and 'supra-personal' character of verbal meaning, which makes it 'a special kind of intentional object' (218). Here the intentional and conventional components of Hirsch's system are designed to meet and complement each other in his redefinition of verbal meaning as 'that aspect of a speaker's "intention" which, under linguistic conventions, may be shared by others' (ibid.). The sharable 'aspect' is the content of the speaker's intention, according to Husserl's 'experience'/'content' distinction (nonverbal and verbal aspects of the intention, respectively).

The 'intentional' (in the phenomenological sense) description of verbal meaning also permits Hirsch to explain the presence of 'unconscious' or apparently unwilled meanings. This is one of the major arguments offered against intentionalism by Beardsley and others: the problem of the undeniable presence in the text of meanings not consciously intended by the author. Where previously such meanings were explained away by Hirsch as being part of the willed type, all the traits of which the author may not have been aware of, they are now seen as belonging 'to the intention as a whole' (1967: 221). Indeed, generic or type boundaries are now somewhat awkwardly superseded by the Husserlian idea of 'horizon' and the interpreter must determine 'the horizon which defines the author's intention as a whole' in order to deduce which of all the possible implications 'are typical and proper components of the meaning' and which are not (ibid.). The horizon concept helpfully delineates the meaning– possibilities, by excluding those that obviously do not 'belong', and it also 'frees the interpreter from the constricting and impossible task of discovering what the author was explicitly thinking of' by exactly replicating his mindset at the time of writing (223). The act of 're-cognition' that Hirsch is asking the critic to perform becomes then a more manageable task. Nevertheless it remains to be asked why Hirsch feels the need to supplement (or even *supplant*) his type theory with a theory that is, first, not his own, and second, reduced and manipulated in such a way as to be closely analogous to the explanation of meaning that he has already offered.

Furthermore, while the phenomenological reduction that Hirsch indulges in here suits the abstractness of his argument, his piecemeal borrowing from Husserl threatens a fall into Husserlian style idealism: it is the object as *experienced* that is the focus, rather than the object in itself. For Husserl all that can be known of the world is our experience or 'intention' of it; compare Hirsch's comment that 'objects *for us* are the only objects we have' (1976: 4). This is not to say that there is no mind-independent reality, only that we cannot know for certain that there is and that, existent or not, a mind-independent reality is of little relevance to us. Through that Husserlian 'epoché' – the method of reduction which performs a kind of mental bracketing off of all but 'pure' consciousness, abstracted from the natural world – objects are reducible to their experienceable properties, that is, to our 'intention' (in the phenomenological sense) of them:

> There are [. . .] not two things present in experience, we do not experience the object and beside it the intentional experience directed upon it [. . .] only one thing is present, the intentional experience, whose essential descriptive character is the intention in question. (Husserl 2001b: 98)

Hirsch himself refers to 'bracketing' as 'a simplified visual metaphor for our ability to demarcate not only a content but also the mental acts by which we attend to that content, apart from the rest of our experience', and it forms the basis of his distinction between meaning and significance, a distinction which mimics Husserl's distinction between the 'inner' and 'outer' horizons of meaning (1976: 4–5). Hirsch's interpreter–critic is primarily concerned with *meaning*, which is variously defined as 'that which is represented by a text; it is what the author meant by his use of a particular sign sequence; it is what the signs represent' and as 'the whole verbal meaning of a text' (1967: 8, 2). This meaning must, however, be distinguished from *significance*, which 'names a relationship between that meaning and a person, or a conception, or a situation, or indeed anything imaginable' (1967: 8). Hirsch asserts the 'natural and universal' character of this distinction, while his detractors decry its 'artificiality' (1976: 3). More pertinently, he elucidates the distinction with a comment on the need to distinguish 'a content of consciousness from its contexts' (ibid.) – a confusing claim as it appears to identify meaning with the content of consciousness, something which he has already denied in his discussion of mental contents and intentional objects, and the differences between them. Such an identification makes him prey to the standard 'inaccessibility' objections which he has tried so hard to avert. In addition, an anti-intentionalist critic could still argue that the object (i.e., the text) in itself, rather than the author's 'intention' of it, is the proper focus of criticism. Interestingly, Ray divides all theories of literary meaning into two clear types: those which figure meaning as 'historically bound *act*', intention-governed, and those which see it as 'permanent textual *fact*', transcending particular volition (1984: 2). Hirsch tends to privilege the view of literary meaning as act, and uses Husserlian phenomenology both to underline the interconnectedness of meaning and consciousness and to evade the objections traditionally incurred by this view. However, like Husserl he fails to shake off his latent psychologism and so is subject to the same accusations of solipsism (the problem of how one transcendental ego is to experience another), essentialism and idealism.

Perhaps this is why Hirsch is careful to emphasize that meaning for him is never *entirely* a matter of consciousness but is always partially constituted by convention. His typological account of meaning detracts from the focus on consciousness, instead figuring meaning as fact (as dictated by generic conventions) and suggests that Hirsch's may fall into that category of 'radical' theory deemed by William Ray to be '[an] attempt to implement both sides of a paradox at once' (1984: 5). Actually, Ray rather perversely denies Hirsch his place in this estimable canon, attacking him for overlooking 'the reciprocity in the relationship between private act and public fact' and favouring 'history over the individual' (90). He further alleges that Hirsch's terminology ('meaning', 'significance', 'validation', etc.) serves to 'reinforce a hierarchy of authority that subordinates individual member to larger class' (97). It is my contention, however, that Hirsch's entire system is in fact an attempt (although not always a successful one) to explicate precisely this 'reciprocity' which is illustrative of the paradoxical nature of literary meaning: the public/private, object/act tension out of which it arises. I would offer the same rejoinder to Meiland's reading of Hirsch (the antithesis of Ray's reading) as an undiluted meaning-as-act theorist, a reading that Hirsch unwittingly facilitates with his characterization of the text as a collection of 'mute signs' (1967: 26). In fact this 'muteness' tag is intended only to reinforce the reciprocal nature of interpretation, where intention and convention are mutually dependent, inextricable. Meiland reads it as a declaration of the overweening power of consciousness in constructing meaning, but this would allow any sign to mean anything the author intended, and Hirsch never claims this. Although he posits that 'there are but two alternatives: either the text represents the author's verbal meaning or it represents no determinate verbal meaning at all', the key word here is *determinate* (234). Hirsch does not claim that the text considered apart from the author's intention has *no* meaning, he merely disputes the relevance of this non-authorial meaning (or meanings) to interpretation. For Hirsch, meaning is a matter *both* of consciousness and of linguistic norms; it is in the mind and in the text.

However, his description of the particularity of intrinsic genre (each text has its own) slightly militates against this dual approach if (and only if) the intrinsic genre is identical with the singular meaning of the text, if what the author 'wills' fails to go beyond a shared category (type) of meaning – and, as we have seen, while Hirsch attempts to argue against such an identification/reduction, he is not entirely successful in doing so. Nevertheless, I would dispute Irwin's contention that meaning is so much a matter of consciousness for Hirsch that linguistic form or the textual presentation of this meaning is irrelevant. Irwin himself sees meaning as entirely a matter of consciousness, and he attacks the phrase 'verbal meaning' as 'inappropriate' because it implies that meaning belongs to the words rather than the creative consciousness which produced them (1999: 58); convention, by contrast, is treated as somehow after the fact – the way that words have come to be used, when really they could be used in any way. This seems to me to be an unnecessary radicalization of Saussure's 'arbitrariness of the signifier' thesis: the *original* arbitrariness of linguistic conventions doesn't preclude them *subsequently* displaying a constitutive function in meaning-production; a reader will read conventionally unless instructed to do otherwise. Furthermore, Irwin's assertion of meaning as entirely a matter of consciousness, part of the 'development and refinement of Hirsch in the author-based tradition initiated by Schleiermacher' that he claims to be engaged in, is compromised by his concession that the reader must look to the text for evidence of the author's likely intention (64). In practice – and Irwin's account does purport to be normative rather than descriptive, that is, to be methods- and goals-oriented – this is a theoretical difference that fails to make a practical difference as the likelihood is that the reader will still have to appeal to the text as the indicator (probably the sole indicator) of authorial meaning.

Further refuting Irwin's claim that meaning, for Hirsch, is entirely a matter of consciousness, Hirsch's more recent contributions to the intentionalism debate present a revised concept of meaning concomitant with his gradual realization that 'meaning is not simply an affair of consciousness and unconsciousness' (1984: 202). The revised definition, which he views as 'a deepening of the concept of meaning', aims to account for future meanings which go beyond the present intention of the author, thereby transcending 'our momentary limitations of attention and knowledge', that is, transcending the contents of consciousness at the time of creation because, after all, man is a creature limited in the extent of his intuitions and mental capacity and the author may not have 'in mind' at once all the possible implications and applications of his text (ibid.). Hirsch later refers to this as 'the inherent fallibility of my mental picture of what will take place' and in doing so he reveals a hitherto unacknowledged scepticism about the viability of the blueprint conception of intention and its usefulness in interpretation (220). At the very least, all of his work can be held to prove the point that the relationship between (textual) meaning and the mental must be open to continual re-interpretation.

Poulet's phenomenology of reading

A more thoroughgoing use of phenomenology in the service of literary criticism occurs in the work of Georges Poulet and, in contrast to Hirsch, his focus falls more squarely upon the reader and the literary 'object' itself. This literary 'object' becomes, for Poulet, an *intentional* object, 'given' only in the perception and/or intentional understanding of the author and the reader, who together constitute it *by* intending it (that is to say, in their shared consciousness *of* it) but who are themselves, in turn, constituted as subjects through their intentional awareness of this literary object. Meaning, for Poulet, is *entirely* a matter of consciousness, but 'consciousness' is here expanded to incorporate the objects of perception, as well as the acts of perceiving them. Specifically, meaning for Poulet is the meeting of two consciousnesses, that of author and reader: in the act of reading I meet 'the consciousness of another', one which, significantly, is 'open to me' and which 'allows me, with unheard-of licence, to think what it thinks and feel what it feels' (1969: 54). There is undoubtedly an utopian dimension to this for Poulet, as the barriers between one mind and another, and between the mind and its objects are dismantled to allow a more direct communion between author and reader and between work and reader than might hitherto have been imagined. We move, he explains, beyond the traditional triangulation of author–text–reader and experience, 'a moment when we are in genuine fusion, when there are no longer *three*, when there are no longer *two*, when there is only *one*' (Macksey & Donato, 1970: 84).

If this seems to suggest the immateriality of 'the text itself' (that stable and solid object of New Critical lore, the 'verbal icon') – which, according to the logic of intentionality detailed earlier, need not attain actual existence, but merely existence/appearance in the consciousness of author and reader (presence to mind, as it were) – then this complies with Poulet's invocation that books must in fact transcend their status as material objects and become something else, something more: 'Books are objects. On a table, on bookshelves, in store windows, they wait for someone to come and deliver them from their materiality, from their immobility' (1969: 53). Their 'immobility', it appears, is a consequence, a symptom of their materiality. Mobility – and, he implies, meaningfulness – comes only with the addition of some animating consciousness or intention; books live only when someone, some consciousness, is intent upon them. Although the book may begin as 'an object among others, residing in the external world' with a 'material reality', it becomes, crucially, an 'interior object' and its significations 'become images, ideas, words', which Poulet describes as 'purely mental entities' (54, 55). The transition or transmutation is a significant one, for 'in order to exist as mental objects, they must relinquish their existence as real objects' (55). Nevertheless, the book does *begin* life as a material entity, so we might ask to what extent the 'interior object' that it becomes is supervenient upon the material object that it first is, as it waits for the reader who will facilitate its transformation. Notably, this movement (from the material to the mental) is quite at odds with the Formalist conception of the text itself as public object, possessed of a concrete reality, existing in the world beyond whatever intentions its author might have for it; the latter can be rethought as a movement from the mental (the authorial blueprint) to the material (the public object which the critic can then take charge of).

With the transcending of this limiting and awkward materiality comes the concomitant dissolution of the merely material boundaries between read text and reading subject who are now linked, as object and subject, by this relationship of intentionality:

> [A book] asks nothing better than to exist outside itself, or to let you exist in it. In short, the extraordinary fact in the case of a book is the falling away of the barriers between you and it. You are inside it; it is inside you; there is no longer any outside or inside. (54)

This 'falling away' of barriers is, in effect, the dissolution of the boundary between inside and outside as the materiality of the text is elided; it is no longer possible to insist on distinctions between 'internal' and 'external' evidence in interpretation, as Wimsatt and Beardsley wished to do; the very logic of inside and outside is overthrown. Furthermore, the imaginative immersion in the work that the reader enjoys is a step beyond Coleridge's willing suspension of disbelief, as the book as material object 'disappears' and becomes instead:

> A series of words, of images, of ideas which in their turn begin to exist. And where is this new existence? Surely not in the paper object. Nor, surely in external space. There is only one place left for this new existence: my innermost self. (ibid.)

Here Poulet gives us a subject-centred theory of meaning where the reader does not so much usurp the author as become one with him and with the piece of literature which brings them together. It is the reader's consciousness which is pre-eminent in the construction of the book as literary 'work', that is in the construction of the series of words, images and ideas which it has become, because the subjective consciousness is the site of the playing out of these images and ideas. Poulet hints, however, at the passivity of this process, and its nature as process and transformation. As he expresses it, 'I deliver myself, bound hand and foot to the omnipotence of fiction. [. . .] I become the prey of language. There is no escaping this take-over' (55). We should not be led, however, into the mistaken assumption that this is an agonistic process, comparable to Blanchot's literary work as 'violent, impersonal affirmation', which 'annihilates' the writer and is 'free of' the one who reads it (1982: 193, 201). In fact, Poulet's reader is 'freed from [their] usual sense of incompatibility between [their] consciousness and its objects' and 'the opposition between the subject and its objects has been considerably attenuated' (1969: 55); by extension, that reader is now free of the anxiety which this impression of incommensurability may bring about. The expansion of subjective consciousness implied in the statement 'everything has become part of my mind' is a reassuring, consolatory one, in Poulet's representation of it (ibid.).

The subject itself does not remain unmoved by this relation with the work, however, as reading effects an alteration of consciousness, allowing you to experience as your own the thoughts of another, and it thus effects a transformation of the reading subject himself whose 'I', whose consciousness is 'on loan to another', which 'other' 'thinks, feels, suffers, and acts within me' (57). As Poulet proceeds to clarify: 'Reading [. . .] is the act in which the subjective principle which I call *I*, is modified in such a way that I no longer have the right, strictly speaking, to consider it as my *I*' (ibid.). The stability of self-identity is thus tested and shaken in the act of reading. What is more, this 'other' whose consciousness invades that of the reader is not, in any straightforward sense, the author, certainly not if we mean by that the historical author: a mass of biographical details external to the work. Although not utterly unrelated to it, as 'every work of literature is impregnated with the mind of the one who wrote it', the author is displaced here by both work and reader (58). Poulet's attitude to the author is made clearer in responses to questions, following the delivery of his original paper. The transcript of this question and answer session is included in Richard Macksey and Eugenio Donato's *The Structuralist Controversy* (1970), where Poulet's 'Phenomenology of Reading' paper is reprinted. When asked specifically about the role of the author in the act and process of reading, Poulet asks for clarification: 'Is your question related to the intentionality of the author himself, or to something else?' and his interlocutor (Charles Singleton) elaborates:

> But now his intention, his intentionality is objectified in this work in such a way that you, not the author, but the reader, can be in communication with it and we follow you there. Now replace that *I* with the authorial *I*. (79)

Singleton is actually asking a question about the author *as reader of his own work*, but Poulet's more general reply is telling, in what it gives away about his attitude to the traditional supremacy of the author and to the inside/outside distinction that has proved so central to our discussion of intention so far. He says:

> I would not replace the *I* within the work by an *I* which would be outside the work, and if we put the name of Dante for instance as the name of the author in question, I would say that the only Dante with whom I would be preoccupied would be the Dante within the work, who is there in two ways. He is there as the hero of the poem; he is there also as the very intentionality present in the poem, present in the poem in such a way that he cannot be distinguished from the poem itself. I have no need for this other Dante who, it is said, is the author of the *Divina Commedia.* (M&D 1970: 79–80)

There are evident flaws in this reasoning – not least the apparent identification of historical author and protagonist in his description of Dante as 'the hero of the poem'. Nevertheless, he appears to reiterate here some version of Wimsatt and Beardsley's notion of 'effective intention' or intention in the text, as the text is, in some manner, infused with this 'intentionality' of the author. Yet, crucially, 'the text' for Poulet is something quite different to Wimsatt and Beardsley's 'verbal icon' for, as we have seen, it is transformed by the act of reading to become something significantly immaterial and subjective; it is precisely this different understanding of what a work is that facilitates this more comprehensible notion of authorial intentionality *in* the text. Incidentally, when pressed on the question of whether the author can ever 'be a reader of his own book', Poulet exclaims: 'Why not, why not! When he has forgotten that he has written it, there must be, I think, a way of perfect detachment in which the author can read his own stuff' (1970: 80). In order to become a 'reader' then, the author must forget that he is the author of what he is reading and thus forgo the traditional position of authority and pre-eminence allotted to authors.

This grants the reader a new kind of agency, even though they are not the *originating subject* (source) of the work that they are engaged in reading and even though they are in some important sense *subject to* that work, which makes them its 'prey'. For, in tacitly challenging the author's authority, Poulet challenges also that author's ownership of his ideas. As he argues, 'Ideas belong to no one. They pass from one mind to another as coins pass from hand to hand', therefore:

> Whatever these ideas may be, however strong the tie which binds them to their source, however transitory may be their sojourn in my own mind, so long as I entertain them I assert myself as subject of these ideas; I am the subjective principle for whom the ideas serve for the time being as the predications. (1969: 56)

This *assertion* of the reader, of reader agency and the reader's subjectivity counteracts any impression that the reader's consciousness is simply a blank space, awaiting colonization, despite the extent to which they are 'inhabited' or 'possessed' by the world of the work. (If such terminology suggests a mystical dimension to the reading process then arguably this is deliberate on the part of Poulet.)

Furthermore, 'the work' (as distinguished from the more material and immobile 'book', a distinction which Blanchot perhaps makes with more clarity[1]) itself displays a consciousness and agency beyond that imbued in it by some author. As he begins to read, Poulet realizes that 'what I hold in my hands is no longer just an object, or even simply a living thing. I am aware of a rational being, of a consciousness; the consciousness of another, no different from the one I automatically assume in every human being I encounter' (54). The work acts upon the consciousness of the reader as if it were *itself* a mind, capable of mental 'acts', capable of intending objects: 'It is the work which traces in me the very boundaries within which this consciousness will define itself' (58). This, then, is rather different to Hirsch's scenario, where the reader or critic appeals to the authority of the author–creator to determine such boundaries or at the very least looks for the consciousness of the author manifest in the text. As Poulet explains, 'It is the work which forces on me a series of mental objects and creates in me a network of words, beyond which, for the time being, there will be no room for other mental objects or for other words' (58–59). In other words, the work places before the reader a number of fictional objects and events and characters and scenarios (briefly: phenomena) – no less real to him, no less intensely experienced or apparent, for their fictionality – and the reader experiences these phenomena, has mental 'presentations' of them which do not differ in kind from the mental presentation of objects not found within the pages of books. And in so acting upon his mind, the work becomes, significantly: 'A sort of human being [. . .] a mind conscious of itself and constituting itself in me as the subject of its own objects' (59).

The language of subject and object remains, the language of the phenomenological model of consciousness, although the parameters of these terms have shifted and likewise their applicability. The work becomes the subject; only subjects, remember, are capable of agency, only subjects are characteristically conscious and intentional. The work – a series of images, an ideal rather than a material object in Poulet's conception of it – is thus seen to exhibit an animation comparable to human consciousness, characterized by intentionality, by its 'directedness' upon intentional objects (in this case, both the consciousness of the reader which it targets and occupies, and the objects and events depicted in the work, which it 'intends'). The work, says Poulet, 'thinks itself' to a certain extent, there is a 'conscious subject ensconced at the heart of the work', inherent in it, and this is not to be readily identified with the subject of the historical author–creator, whom it surpasses, whose dictates and imaginings it exceeds: 'The consciousness inherent in the work is active and potent; it occupies the foreground; it is clearly related to its *own* world, to objects which are *its* objects' (ibid.).

Radical as this attribution of subjective consciousness to a literary work is, this is still subjectivity expressed through the objects which it intends and in this sense not an isolation of the subjective domain (again, resisting any accusations of idealism); subject and object still reciprocally constitute each other, are still interdependent. However, it would be possible to argue that the objective status of those objects intended by the work itself is further weakened by their fictionality, making them much more obviously the projections of some subjective consciousness (whether author's or reader's or work's) which demonstrate no independent existence. Accordingly, at the close of his article, Poulet does gesture towards what he terms 'subjectivity without objectivity' (68). He insinuates that although there is, within a literary work, a 'mental activity' which is 'profoundly engaged in objective forms' there is, in addition, and 'forsaking all forms':

> A subject which reveals itself to itself (and to me) in its transcendence over all which is reflected in it. At this point, no object can any longer express it, no structure can any longer define it; it is exposed in its ineffability and in its fundamental indeterminacy. (ibid.)

This is work as pure intention, pure self-presence, even before the appearance of objects to be intended or formal structures for the realization of this intentionality – yet this 'before' is misleading, because even this supposedly pure intentionality is dependent upon the subjective consciousness of the reader, who brings the work *qua* work (which is, of course, the work-as-consciousness) into existence. The work cannot exist as anything more than base matter, words on a page, cannot express this 'ineffability' of which he speaks, without the animating consciousness of the reader – so it is perhaps best described as intersubjective.

Poulet's reading subject may not 'intend' the meaning of the text in the same way as Hirsch's originating authorial subject, but he still provides the necessary basis of self-presence and the injection of agency for meaning to 'occur', to be apprehended, knowable. What is interesting is the way that Poulet moves towards a vision of the comparable agency of the literary work itself, in the way it initially 'solicits' our attention and subsequently enacts a kind of imaginative colonization of the readerly consciousness. In doing so he inches towards the conception of a transcendental subjectivity which is 'anterior and posterior to any object' (1964: viii). So subjective consciousness is once more central to the enterprise of reading and understanding literature, but the very nature of the subject and the mental has been altered with the figuring of the text itself as a form of consciousness or conscious agency. Furthermore, in the case of the reading subject, it is a colonized consciousness, with all the ambivalence that this implies – an ambivalence which Poulet, in the *jouissance* of his account of the phenomenology of reading, fails to elaborate and assimilate.

Ingarden and Iser: Concretization and the role of the reader

Alternative but comparable accounts of the phenomenology of reading are to be found in the work of Roman Ingarden (whose landmark book *The Literary Work of Art* first appeared in German in 1931, but was not translated into English until 1973) and Wolfgang Iser. In both cases, the relationship between work and reader is crucial (with Iser placing slightly more emphasis on the latter than Ingarden), and in both cases the 'phenomenological' approach offers a means of characterizing and explaining both the intentional nature of the literary work and that work's liminality vis-à-vis the act/object, subject/object divides.

If Poulet is primarily concerned with questions of the readerly consciousness, then Ingarden is, by contrast, more anxious to avoid allegations of psychologism, arguing in fact that 'the dangers' – of psychologism and, by extension, subjectivism – 'have an incomparably greater significance than the relatively unimportant matter of literary theory' (1973b: 359). If literary works are merely subjective 'concretizations' in the consciousnesses of their respective readers, warns Ingarden, then it is but a short step to the suggestion that scientific works are similarly constituted, and thus 'that *intersubjective* knowledge is impossible' (ibid.). The major focus of *The Literary Work of Art*, then, comprises his attempt to establish the objective character of the intentional object (here, the literary work) itself, its essential qualities. This proves to be, from the outset, a fraught exercise, as Ingarden initially ponders whether the literary work is a 'real' or an 'ideal' object and at once concedes that 'it is not immediately clear what a literary work actually is' (10). In the course of his argument it becomes apparent that the literary work is not, in any straightforward sense, either ideal or real. Levin's explanation helps to elucidate Ingarden's dual approach to the problem by attributing to him the view that 'every entity which is to count as a literary work, however innovative or however experimental it may be, exhibits a certain objective structure, the rules of which we can articulate and indeed decisively ground in the unquestionable sense of our literary experience' (xvi). This description highlights Ingarden's goal of revealing and elaborating the *rules* which govern the make-up of all literary works (and thus his belief that such rules exist and can be elaborated), but it also shows Ingarden's method to be based on an investigation of both *objective structure* and (subjective) *literary experience* and, by extension, of the vital reciprocity of the two which makes the literary work what it is.

As far as the inception of a literary work is concerned, Ingarden insists upon a distinction between 'the basis of the *coming into being* of the literary work and the ontic basis of its existence (i.e., after formation)', claiming that 'we have already found the former in the subjective operations the author executes when forming the work' (360). These 'subjective operations' are not of primary interest to him and his account is therefore quite at odds with the idea of re-cognition (of authorial intention) that we find, for example, in Hirsch. In fact Ingarden argues that:

> No one would want to identify the concrete psychic contents experienced by us during the reading with the already long-gone experiences of the author. [. . .] [T]he attempt to identify the literary work with a manifold of the author's psychic experiences is quite absurd. The author's experiences cease to exist the moment the work created by him comes into existence. (14)

This refusal to identify the work (and its meanings) with the mental content of the author at the time of writing clearly distinguishes Ingarden's approach from any kind of naïve authorial intentionalism. Any attempt to access or reconstruct these 'authorial experiences' would, then, be misguided; but Ingarden doesn't explain the ontology of the work via the contents of the reader's consciousness either, claiming similarly that 'the view that the literary work is nothing but a manifold of experiences felt by the reader during the reading is also altogether false and its consequences absurd', for then 'every new reading would produce an entirely new work' (15). (His sentiments here rather interestingly anticipate Wimsatt and Beardsley's intentional and affective fallacies.) These assertions serve as his argument against 'the psychologistic conception of the literary work' and reveal the essential reciprocity between (and inextricability of) textual structure and readerly concretization in the construction of a literary work of art (ibid.). In order to elucidate this relationship, it is necessary to investigate the meanings of both structure (or, more accurately, schema) and concretization in Ingarden's work, and in Iser's subsequent borrowings from his phenomenological forebear.

Iser's own reading of the relationship between the text and its realization is useful here, for he concludes that:

> From this polarity [of text/realization, or artistic/aesthetic pole] it follows that the literary work cannot be completely identical with the text, or with the realization of the text, but in fact must lie halfway between the two. The work is the point of convergence, since it is located neither in the author's psyche nor in the reader's experience. (2006: 15)

As a 'point of convergence', the work (as distinct from the text) is neither ideal nor real, but is instead 'an intentional object, whose component parts function as instructions, the execution of which will bring the work to fruition' (ibid.). Ingarden himself describes the work as 'purely intentional', qualifying this with the explanation that:

> It exists as an ontically heteronomous formation that has the source of its existence in the intentional acts of the creating conscious subject and, simultaneously, the basis of its existence in two entirely heterogeneous objectivities: on the one hand, in *ideal* concepts and ideal qualities (essences), and, on the other hand [. . .] in real 'word signs'. (1973b: 361)

He therefore proceeds, in quite laborious detail, to set out the different 'strata' which make up a literary work, and which function (as Iser's reading, given earlier, indicates) as 'instructions' for the reader. Ingarden's literary work is made up of four strata which are 'necessary for every literary work if its internal unity and basic character are to be preserved':

> (1) the stratum of *word sounds* and the *phonetic formations* of higher order built on them; (2) the stratum of *meaning units* of various orders; (3) the stratum of manifold schematized *aspects* and aspect continua and series, and, finally, (4) the stratum of *represented objectivities* and their vicissitudes. (30)

He describes his analysis as '[cutting] a cross-section through the structure of the literary work', although this cross-section 'does not permit us to apprehend the total nature of the literary work of art' (304). This is because part of that nature, at least, depends upon the concretization of the work by the reader, which is necessary because the work itself is 'schematic', and contains '"gaps", spots of indeterminacy' as well as a kind of 'potentiality' which awaits realization or discovery (331). The 'concretizations' help constitute 'the mode of appearance of a work, the concrete form in which the work itself is apprehended' (332). The language of the Husserlian 'horizon' is evident here in Iser's explication of Ingarden's account of sentence sequence:

> The semantic pointers of individual sentences always imply an expectation of some kind. As this structure is inherent in all intentional sentence correlates, it follows that their interplay will lead not so much to the fulfilment of expectations as to their modification. Each individual sentence correlate *prefigures a particular horizon,* but this is immediately transformed into the background for the next correlate and must therefore necessarily be modified. (2006: 17; my emphasis)

This notion of 'prefiguring' indicates the ways in which the very structure of a literary work (the structural relationships between its sentences) serves to delimit the possible concretizations of it, without thereby determining those concretizations; in line with my previous discussion of Hirsch and Meiland,[2] the textual schema can be seen as ruling out more than it rules in. The language of continual 'modification', also reveals this to be a temporally extended *process* where the meaning-horizon is subject to continual revision, as are the reader's expectations.

In setting out what is involved in the reading process, Ingarden is keen to distinguish between the work itself (as adumbrated earlier, its schema) and the reader's concretizations of it, and between those concretizations and 'the subjective operations and [. . .] psychic experiences we have during the reading', our acts and experiences of 'apprehension' (1973b: 332, 335). It is concretization by the reader which completes the work (at least temporarily), by filling those 'spots of indeterminacy' in a work and by turning its 'potentialities' into 'actualizations' (337, n9). However, this requires some effort on the reader's part, who must adopt 'an attitude of pure beholding' where they are attentive to the complexity of the work itself and its 'represented objectivities' and positively inattentive to their own surrounding reality (335). In this way Ingarden depicts a 'communion' between work and reader not unlike that described by Poulet, yet works to shield himself from the accusation that his literary work is a merely psychic phenomenon. Throughout, the 'intentional' character of the work is maintained, in positioning it as something which emerges out of the relation of text and reading, so that description of it (by Iser) as a 'point of convergence' can now be understood as a comment about the necessarily inextricable, mutually constitutive nature of act and object which phenomenology identifies and which it terms 'intentionality'.

There are issues which might give us pause, however, when it comes to the adoption of Ingarden's model of literary intentionality. As with speech act theory, it is possible to ascertain here a reliance on a problematic distinction between literary language and ordinary language. While the 'declarative sentences' of a scientific work, writes Ingarden, 'are genuine *judgements* in a logical sense, in which something is seriously asserted and which not only lay claim to truth but *are* true or false', those of the literary work, by contrast, 'are not pure affirmative propositions, nor, on the other hand, can they be considered to be seriously intended assertive propositions or judgments' (1973b: 160). Later he writes that 'in a literary work there are only quasi-judgmental assertive propositions of various types' (171). As Iser points out, however, this raises the question (which remains unanswered by Ingarden) of how we tell the difference between texts which consist of 'assertive propositions' and those which consist of only 'quasi-judgmental sentences' (2006: 18).

In addition, we might find fault with Ingarden's insistence on the unity of the text, his proto-structuralism, which raises the question of whether a work can be described as 'intentional', without the assumption of such a unity. Yet Ingarden insists also on the work's indeterminacy, and hints at the contingency of any concretizations of it, and in doing so he gestures towards a view of the text that could more accurately be described as post-structuralist. The tension in his work between this investment in unity and a positing of the literary work's indeterminacy in fact arises because he's keen to avoid the outcome where the work could mean anything we want it to mean, but also anxious to present reading as something other than an entirely passive process. Wolfgang Iser, in his reading of Ingarden, appears to reiterate the emphasis on unity, but in doing so he also suggests a way out of the conundrum. For while the interaction and intersection of the intentional sentence correlates ultimately '[gives rise] to the semantic fulfilment at which they have aimed', this fulfilment, 'takes place not in the work but in the reader' who 'activates' or 'concretizes' the meaning of the work (17). The work then achieves its unity only through the action of and in the moment of its concretization, not *in itself*. Iser's development and alteration of Ingarden's philosophy, as part of his own analysis of the reading process, is worth considering here, so I shall turn to that now.

Writing in 1978, Iser states his aim to move beyond what he sees as the limiting 'norms' of interpretation as traditionally practised, and among these he cites the focus upon authorial intention, arguing that:

> So long as the focal point of interest was the author's intention, or the contemporary, psychological, social, or historical meaning of the text, or the way in which it was constructed, it scarcely seemed to occur to critics that the text could only have a meaning when it was read. (1978: 20)

This, for Iser, is the crucial point about literary meaning: that it arises out of a reader's interaction with a text rather than being something put into the text by its author or something inhering in its construction; the reader doesn't simply respond to an already-present meaning, he actually helps to bring the work to fruition and thus displays a greater agency than that implied by a kind of Hirschian re-cognition of authorial meaning.

In 'The Reading Process: A Phenomenological Approach', Iser takes Ingarden's idea of concretization as his starting point, but contributes his own terminology to the debate when he writes of the 'convergence of text and reader' as always 'virtual' and the work which emerges from such a convergence (the work which just *is* this convergence) as 'virtual' and 'dynamic' (1974: 275). This 'work' is distinguished by Iser (again, taking the lead from Ingarden) from the *text* and in fact he figures text and work as poles: while the text (the particular collection of sentences) is created by the author, the literary work is realized by the reader; the former is 'artistic', the latter 'esthetic'; the work 'is more than the text, for the text only takes on life when it is realized' (274). In this last statement we can hear echoes of Poulet, in his description of the work's deliverance from its immobility, its materiality, although Iser himself does not use quite this terminology. The virtuality of the work is 'inevitable' precisely because of that work's location at/as the convergence of two disparate entities, for in that convergence the work becomes more than the sum of its parts, not being reducible to either 'the reality of the text' or 'the subjectivity of the reader'; such a virtuality lends to the work a quite unique 'dynamism' (1978: 21). Like Ingarden, Iser is at pains to explain the finer points of this interaction between text and reader which brings about the work. While this is to some extent a subjective process, for 'the realization is by no means independent of the individual disposition of the reader', yet it is also reciprocal, for this 'individual disposition of the reader' is also 'acted upon by the different patterns of the text' (1974: 274, 275). In *The Act of Reading*, Iser is even clearer in his insistence that this interaction between the 'reality of the text' and the 'subjectivity of the reader', between 'structure' and 'recipient', is not a relationship of 'transmitter and receiver', for 'in literary works [. . .] the message is transmitted in two ways, in that the reader "receives" it by composing it' (1978: 21). Both reader and text are active here, they act upon each other (interact) to produce the work.

It is in this implied agency of the text, which 'offers' the reader 'various perspectives', ways of relating the 'patterns and the "schematised views" to one another', that we can detect a kind of intentionality at work on the part of the text itself (1974: 275). Nevertheless, Iser himself shies away from such a thesis, suggesting still that it is the reader who 'sets the work in motion': 'reading causes the literary work to unfold its inherently dynamic character' (ibid.). It is reading which is the main focus of Iser's enquiry, yet we might ask in what that *inherently dynamic* character of the work consists, for in that term 'dynamism' is implied change, activity, process, a force which is somehow stimulated and structured by the text itself (part of its structure and its structuring operations), even if it requires the reader to actualize it. At the very least, the use of metaphors of movement and change is interesting, for the text is no longer a 'verbal icon', something static and determinate, but rather, in its meaning potential, subject to a temporal unfolding; arguably it is not only the act of reading which is a 'process' here; the work too becomes more process than semantically autonomous icon, and this must have a knock-on effect for our understanding of literary meaning and intention. Yet Iser's emphasis is determinedly upon the temporality of the reading process itself (on reading as process), on the 'active interweaving of anticipation and retrospection' involved in reading (282). He pays much less attention to the temporality of the work itself: its alteration over time (rather than simply its alteration *for the reader* or in the perception of the reader).

The interaction of text and reader, to the extent that it can be situated, occurs in the imagination: the text stimulates the imagination of the reader, and the reader imaginatively completes the work. Iser writes admiringly of Laurence Sterne's understanding of a literary text as 'an arena in which reader and author participate in a game of the imagination' (275). So reading is an 'active and creative' process – filling in gaps ('the "unwritten" part of the text'), establishing connections, finding meaning where this may not be obvious (ibid.). Yet the text too has an active role to play, for 'the written text imposes certain limits on its unwritten implications in order to prevent these becoming too blurred and hazy', so, as with Ingarden's approach, the text is held to rule out certain implications, certain meaning-possibilities and thus the work cannot be simply a product of the reader's imaginative engagement with it (276). The use of 'implications' here is, again, reminiscent of the Husserlian horizon of possible meanings as invoked by both Hirsch and Ingarden. More positively, the text also creates 'expectations' which are never quite fulfilled, so it functions productively, creating its own meanings, not merely negatively, by setting limits to the reader's imaginative engagement with it and uses of it.

In fact, Iser's representation of the semantic open-endedness of the work anticipates the deconstructive accounts of textual meaning that I will go on to discuss in the next chapter. Iser's textual 'implications' must then be distinguished from those of Hirsch, given the latter's investment in notions of textual determinacy and interpretative validity; indeed, the former sets out to oppose interpretative monism in its various forms, while yet defending his approach against charges of an out-and-out relativism or subjectivism. As he describes it, 'the meaning of a literary text is not a definable entity but, if anything, a dynamic happening' – again, this shift from *entity* to *happening* is significant, situating the work at the convergence of object and act, rather than identifying it with either – and the text's polysemantic nature is such that the critic must identify its 'potential meanings', rather than imposing a single meaning upon it; the critic must also be aware that 'the total potential can never be fulfilled in the reading process' (1978: 22). It is in this sense that the work is inexhaustible, for it always necessarily exceeds any particular reading of it. The advantage of such 'indeterminacy' is that it facilitates the reader involvement and interaction that Iser is detailing and which, it is implied, goes some way towards explaining our continuing fascination with literary works of art and their value for us. Iser continually emphasizes the ways in which meaning is *not* fixed, highlighting the inevitability of 'continual modification' in reading, and suggesting the essential openness of the work, which can take on different meanings with different/new readings (by the same reader or by different readers) (1974: 278). In fact there is an evaluative slant to Iser's endorsement of this 'conditional modification' of the meaning possibilities of a work (a modification that is both a quality of texts and an outcome of the reading process), as he claims that 'expectations are scarcely ever fulfilled in truly literary texts': the 'truly literary text', then, is one that is sufficiently ambiguous and complex that it gives rise to multiple readings, and is to be distinguished from one which is too 'expository'; such a text becomes 'didactic' and, it is implied, lacks literary value (ibid.).

If Iser is not working towards the establishment of certain objective criteria of interpretative validity as Hirsch is, he still acknowledges the susceptibility of his account of the reading process to allegations of subjectivism and seeks to avert these. As he concedes, 'a reader-oriented theory is from the very outset open to the criticism that it is a form of uncontrolled subjectivism' (1978: 23). His detailed analysis of the process of meaning composition by the reader therefore lays stress on the fact that 'however individual may be the meaning realized in each case, the act of composing it will always have intersubjectively verifiable characteristics' (22). While Poulet muses upon the meeting of minds of author and reader, and Ingarden concentrates largely upon the structure of the text, its various strata, Iser attends to the reader's agency in interacting with those strata, an activity which he considers to be sufficiently conventional and structured, sufficiently beholden to the object of its enquiry that it cannot be held to be 'subjective', strictly speaking. Yet, in placing his emphasis so firmly upon the reader, Iser does introduce this element of subjectivism, for he concedes that, although 'acts of comprehension are guided by the structures of the text', yet 'the latter can never exercise complete control' (24). The structures of the text do not *determine* the possible responses to it, they merely *guide* these responses. If this complex and subtle negotiation between objectivist and subjectivist accounts of literary meaning is not always successful, it does at least seek to provide a workable explanation of the ontological peculiarity of literary texts and our engagement with them.

Nevertheless, in so doing, phenomenological theories of literature such as Iser's are prone, like the speech act theories considered in the last chapter, to too facile an apprehension of the distinction between literary texts and other kinds of utterance figuring them as, perhaps, more peculiar than they really are. Iser follows Ingarden in arguing that fictional texts lack 'the total determinacy of real objects', because they 'constitute their own objects', rather than '[copying] something already in existence' (24); elsewhere he argues that they 'do not correspond to any objective reality outside themselves', accepting without question the idea that normal patterns of reference are suspended in the case of sentences in fictional works and that the world/work distinction can be unproblematically drawn (1974: 276). As a consequence, he claims, the sentences in literary texts do not consist of statements; this would 'be absurd, as one can only make statements about things that exist' (277). Instead, the sentence in the literary text 'aims at something beyond what it actually says' and 'it is through the interaction of these sentences that their common aim is fulfilled' (ibid.). (We might respond that even the most banal sentences employed in everyday conversation can aim at something 'beyond' what they actually say, that their interaction is vital to their overall meaning and, furthermore, that statements in literary texts are still statements, whatever their altered conditions of reference.[3]) Where Iser's account of literary sentences differs from the accounts offered by, say, John Searle or Mary-Louise Pratt is in its comparison of those sentences to Husserl's 'pre-intentions, which construct and collect the seed of what is to come, as such, and bring it to fruition' (ibid.). The crucial shift here is from thinking of literary sentences as expressions of authorial intention, to thinking of them as being (pre-) intentions in their own right.

The emphasis on sequence, pattern and structure here is also interesting, for it is via these structural elements that the text appears to display some agency of its own in helping, along with the animating consciousness of the reader, to bring the work to fruition. Those structures/patterns, the interaction of different elements and so on, do not absolutely determine the meanings of the final work, but if we view them as 'pre-intentions' they certainly have a role to play, in forming 'expectations' (278). Significantly, the structures of which Iser speaks possess the character of 'two-sidedness', being both 'verbal' and 'affective': 'The verbal aspect guides the reaction and prevents it from being arbitrary; the affective aspect is the fulfilment of that which has been prestructured by the language of the text' (1978: 21). In this way Iser evades allegations of the arbitrariness of emotional response (the affective fallacy), while allocating a key, agential role to language itself, explicable now via the phenomenological idea of the pre-intention. This takes us some distance from the idea that the text is a vehicle for authorial intentions which serve to fix its meaning. Rather, 'the literary text activates our own faculties' (1974: 279).

Yet, for every assertion of the literary text's active and 'activating' role, Iser includes several more attributing causal power and ultimate agency to the reader:

> The reader [. . .] causes the text to reveal its potential multiplicity of connections. These connections are the product of the reader's mind working on the raw material of the text, though they are not the text itself – for this consists just of sentences, statements, information, etc. (278)

Such statements can perhaps be qualified, however, by a consideration of what exactly Iser means by 'the reader'. His particular conception of the reader is important because it serves as one of the ways in which he attempts to sidestep charges of psychologism and subjectivism, by distancing his own conception from both ideal and historicized conceptions of the reader. Thus, he criticizes the notion of the 'ideal reader' on the grounds that such a reader, 'would have to be able to realize in full the meaning potential of the fictional text', which Iser has already suggested is an impossible task, and takes issue also with Riffaterre's 'superreader', Fish's 'informed reader' and Wolff's 'intended reader' (1978: 29, 30). In assessing the utility of the last of these, Iser agrees, 'that there is a reciprocity between the form of presentation and the type of reader intended' (33). Developing this point, he proceeds to outline his own theory of 'the implied reader', in the following terms:

> He embodies all those predispositions necessary for a literary work to exercise its effect – predispositions laid down, not by an empirical outside reality, but by the text itself. Consequently, the implied reader as a concept has his roots firmly planted in the structure of the text; he is a construct and in no way to be identified with any real reader. (34)

Such a conception suggests that Iser's emphasis on the reader does not work to contradict his positing of a more active, meaning-activating text, because the reader here emerges from the text, as its 'construct'. That view of the text as active, constructive, intentional (in a phenomenological sense) and as somehow containing the conditions of possibility of its reader (and its reading) is further reinforced by Iser's claim that 'the concept of the implied reader designates a network of response-inviting structures, which impel the reader to grasp the text' (ibid.). The concept of the implied reader incorporates a characterization of that reader as both 'textual structure' and 'structured act': part of the text and responsible for the act of concretization which realizes it, transforming it into a literary work proper (35). Such a characterization ties in to our discussion of intention, as Iser goes on to assert that 'textual structure and structured act are related in much the same way as intention and fulfilment, though in the concept of the implied reader they are joined together in the dynamic process we have described' (36). Once again, it is a phenomenological understanding of intentionality which permits this bringing together of object and act, as the text in some way 'intends' its reader – and vice versa – and they exist in a kind of productive (dynamic) and irresolvable tension. In turn, this allows Iser to maintain that, although '[the] response to any text is bound to be subjective, [. . .] this does not mean that the text disappears into the private world of its individual readers', for 'the process of assembling the meaning of a text is not a private one', and, we might add, if the reader is considered as one of the text's 'implications', then he is not 'private' in the way that such an accusation assumes (49).

Indeed, Iser goes so far as to claim that 'the suggestion that there are two selves is certainly tenable, for these are the roles offered by the text and the real reader's own disposition' and, of these, 'the role prescribed by the text will be the stronger, [although] the reader's own disposition will never disappear totally' (37). This characterization of the text as a 'self' is reminiscent of Poulet's description of the text as 'a sort of human being [. . .] a mind conscious of itself and constituting itself in me as the subject of its own objects' (1969: 59).

The effects on the subjectivity of the reader of this meeting of text and reader are evident in Iser's explanation of what he calls the 'overdetermination' of the text. A literary text is by (his) definition, 'overdetermined', which is to say that it gives rise to multiple possible meanings, becoming more open rather than more closed as it proceeds and as 'the predictability of the individual parts of speech' is reduced – the opposite to what happens in 'everyday speech' (1978: 48). It is this 'overdetermination' which, as he explains it, 'enables the reader to break out of his accustomed framework of conventions, so allowing him to formulate that which has been unleashed by the text' (50). In this way, rather like Poulet before him, Iser suggests the ways in which the reader is fundamentally altered by the act of reading, and again stresses the reciprocity fundamental to this act of reading, for 'it is only when the reader is forced to produce the meaning of the text under *unfamiliar* conditions, rather than under his own conditions (analogizing), that he can bring to light a layer of his personality that he had previously been unable to formulate in his conscious mind' (ibid.). The text, then, reveals and releases something in the reader of which he was hitherto unaware. In claiming this, however, Iser does not go as far as Poulet in his troubling of reader subjectivity.

In glossing Poulet's account of reading, Iser paraphrases it as follows: 'in reading the reader becomes the subject that does the thinking. Thus there disappears the subject-object division that otherwise is a prerequisite for all knowledge and all observation' (1974: 292). However, the author is brought back into this process, for 'the strange subject that thinks the strange thought in the reader indicates the potential presence of the author, whose ideas can be "internalized" by the reader', and thus 'consciousness forms the point at which author and reader converge' (ibid.) and 'the work itself must be thought of as a consciousness' (293). Iser notes that this description of the work is qualified by Poulet's injunctions that 'the life-story of the author' and 'the individual disposition of the reader' be excluded from the work and the reading of the work, respectively (292–93). Nevertheless, he is keen to distinguish his own argument from Poulet's, thinking that the latter's arguments 'should be developed along somewhat different lines', suggesting that the 'division' which occurs in reading is one which 'takes place within the reader himself', and contending that 'although we may be thinking the thoughts of someone else, what we are will not disappear completely' (293). This leads to his conclusion concerning 'the dialectical structure of reading' – 'the need to decipher gives us the chance to formulate our own deciphering capacity' (294). So:

> The production of the meaning of literary texts [. . .] does not merely entail the discovery of the unformulated [aspects of the text], [. . .]; it also entails the possibility that we may formulate ourselves and so discover what had previously seemed to elude our consciousness. (ibid.)

While Poulet's reader appears to be in some sense undone by the process of reading and by being inhabited by the consciousness of another, Iser's reader, by contrast, is constituted, realized and expanded by that encounter.

Iser arguably makes advances on Ingarden too, in his willingness to dispense with the latter's clinging to some ideal of textual unity and coherence: while Ingarden expects one sentence to 'flow' from another, confirming rather than frustrating expectations, Iser asserts that 'literary texts are full of unexpected twists and turns, and frustrations of expectations' (279). It is precisely these 'blockages' and 'gaps' which allow the reader to participate and which, by implication, make for a better literary work: in fact he claims that 'it is only through inevitable omissions that a story gains its dynamism' (280). The text's agency (and its facilitation of reader agency) is, then, predicated on its necessary incompletion. The reader subsequently attempts to form the literary work into some more or less coherent whole, 'projecting' onto the text 'the consistency which we, as readers, require'; what Iser calls the 'gestalt', a kind of unified meaning (284). He adds: 'The "gestalt" is not the true meaning of the text; at best it is a configurative meaning' and involves readerly expectations meeting the text; what unity there is, then, arises from the reader's engagement with the text (ibid.). We set limits to what the work can mean for us, although we are also continually reminded of the polysemantic nature of the work, and the ways in which it exceeds the limits that we attempt to impose upon it and frustrates our attempts to make it into a coherent whole. Iser sees a kind of harmony in this process however, as the 'polysemantic nature of the text' and the 'illusion-making of the reader' (which are 'opposed factors') ultimately balance each other out: 'in the individual literary text we always find some form of balance between the two conflicting tendencies' (285). This rather utopian conclusion somewhat militates against the more interestingly agonistic aspects of the reading process as he represents it.

Conclusion

Hirsch's borrowings from phenomenology, although they sit uneasily within the broader framework of his approach, yet indicate the potential for a re-imagining of intention as intentionality or directedness. Such a potential is visible in Georges Poulet's 'phenomenology of reading', which, in contradistinction to Hirschian authorial intentionalism, shifts the focus from author to reader and to the literary work itself. Most radically, Poulet appears to attribute to the work itself a variety of intentional agency, making that work a subjective consciousness in its own right. What this suppresses, however, is the material status of the text, its dogged material existence, its concrete although not determinate inscription, its formal qualities, its object-ness. Instead, the emphasis is on the work as transcendent, immaterial, purely self-present consciousness: a series of ideas or images 'inscribed' only contingently and transiently in the consciousness of the reader, if anywhere.

Ingarden and Iser, more promisingly, offer accounts of the reading process similarly influenced by Husserlian phenomenology, but with a more intricate understanding of the textual strata which help to structure the reader's response to the text, and of the process by which the work (as a set of meanings) is 'configured' out of the meeting of text and reader. In the case of Iser, we find in addition a conception of 'the reader' as textual construct which averts some of the accusations of subjectivism and psychologism by which phenomenological approaches to literature (like intentionalist theories) have always been tainted.

Iser, then, arguably moves us towards some notion of the 'intention' of the text. Indeed, he claims that this intention is the text's 'unformulated part' – which is not to say that it does not exist, merely that it is inexplicit and therefore requires the reading process to bring it out. He claims that:

> It is [the] interplay between 'deduction' [i.e. working out what is there in the text itself, already] and 'induction' [i.e. imaginatively adding to the text] that gives rise to the configurative meaning of the text, and not the individual expectations, surprises, or frustrations arising from the different perspectives. Since this interplay obviously does not take place in the text itself, but can only come into being through the process of reading, we may conclude that *this process formulates something that is unformulated in the text and yet represents its 'intention'*. Thus, by reading we uncover the unformulated part of the text, and this very indeterminacy is the force that drives us to work out a configurative meaning whilst at the same time giving us the necessary degree of freedom to do so. (1974: 287; my emphasis)

This can be read in different ways. On the one hand, Iser is claiming merely that the reading process uncovers the text's 'intention', and certainly he says little here about what that 'intention' comprises (apart from absence, aporia and frustration). On the other hand, a shift of emphasis, legitimated in large part by Iser's conception of the reader, presents us with the more radical idea that *the text's intention works to structure the configurative meaning that the reader produces*.

Running contrary to such speculative re-readings of Iser's work is the fact that, despite his protestations to the contrary, authorial intention still gets a look-in here, for he implies that, in '[organizing] and [reorganizing] the various data offered us by the text', in looking for 'the fixed points on which we base our "interpretation"', we seek to be faithful to the intention of the author: 'trying to fit them together in the way we think the author meant them to be fitted' (288). This sentiment is compounded by his citation of the following lines from John Dewey's *Art as Experience*: 'to perceive, a beholder must *create* his own experience. And his creation must include relations comparable to those which the original producer underwent' (ibid.). Dewey (and subsequently Iser) calls this the 'act of recreation' and it sounds not dissimilar to Hirsch's idea of 're-cognition', despite Iser's insistence that it 'is not a smooth or continuous process, but one which, in its essence, relies on *interruptions* of the flow to render it efficacious' (ibid.). This summoning of the author, notwithstanding Iser's efforts to conceive of the work as a product of the encounter between text and reader, is evident also in *The Act of Reading*, where the 'perspective view of the world' that the text offers is identified with 'the author's' view, and where the novel is therefore described as 'a system of perspectives designed to transmit the individuality of the author's vision' (1978: 35).

If this suggests that phenomenological theories of reading can only take us so far, then it perhaps warrants the move into the final chapter, where my interest will lie in exploring the possibility of a notion of textual agency which does not involve a concomitant denial or suppression of the material, formal aspects of texts – which, tentatively, radicalizes the notion of form so that intention is literally *incorporated* into the text. In addition, it is the centrality of the subjective consciousness – whether predicated of work or reader – which opened phenomenology and its literary critical derivations up to allegations of psychologism and mysticism, and which ultimately hastened its demise. In fact the turn against phenomenology and towards structuralism can be read as a direct result of the former's emphasis on the increasingly beleaguered subject as the origin or locus of meaning, to the exclusion of those material and linguistic structures which could themselves be seen as producing – rather than merely housing or containing – meaning and subjectivity; the phenomenological focus is too often on force to the exclusion of form, when form might itself be forceful. For Poulet in particular, the work only becomes forceful, or active, when it transcends its base form, its dull, object-ive character, and becomes pure force, pure consciousness. What post-structuralist theories of meaning and text may facilitate is first, the possibility of an intention divorced from subjectivity and, second, an understanding of the complex inextricability of form and force in the intentional production of literary meaning; nevertheless, in doing so, their own debts to phenomenology must also be acknowledged.

Notes

1 See, e.g., *The Space of Literature*, trans. by Ann Smock (Lincoln: University of Nebraska Press, 1982), p. 23ff. Also *The Writing of the Disaster*, trans. by Ann Smock (Lincoln: University of Nebraska Press, 1986).

2 See Chapter 1, pages 27–28

3 The question of the referentiality of literary sentences is discussed at length as part of the account of speech act theory and fictionality in Chapter 2.

Chapter 4

Intention after the Subject

Intention is not seen here as necessarily subjective, but it is seen as necessarily semantic. Intention is inseparable from the concept of meaning; any meaning is to some extent intentional. Any language oriented to meaning is at least intentional, precisely by virtue of the fact that it intends meaning. Intention is, therefore, not subjective.

Paul de Man (1986)

Introduction: The disappearance of the subject?

In the first chapter we saw that although Hirsch develops, through his theory of type meanings, a study of genre which is essentially formalist, yet this is paradoxically dependent upon a notion of the extra-textual individual creator who bears the primary responsibility for choosing and employing the meanings which are the tokens of a particular type. It is Hirsch's agent–author who guarantees the 'validity' of any interpretation of a text and this validity implies a kind of truth: something about which we can be right or wrong, something which does not admit of degree or variation (Hirsch's 'horizons' of meaning may hint at permissiveness but they still amount to boundaries between what is and what isn't 'valid'). Additionally, the 'ethics' of Hirsch's approach are indubitably the ethics of authorship: specifically, the rights owed to the author as 'owner' of the text that he has produced, rights which could not be seen to accrue to the text itself, in his view. The viability of literary criticism as a discipline is indissociable from its humanistic value, what it reveals to us, as human subjects, of some other human subject or subjects. If Hirsch's theory is agreed to be broadly expressionist in character, then what is being expressed by any text is the nature and being of the subject writing.

Similarly, the speech acts discussed by Searle and Austin are always attributable, at least in principle, to some singular speaker, who is conceived as the sole originator of illocutionary force – of which there can also be said to be some 'truth': thus, the speaker is *either* promising *or* threatening *or* stating, he decides to which single use the words will be put. As evidenced by his exchange with Derrida, in the pages of *Glyph* and subsequently Derrida's *Limited Inc*, Searle is reluctant to concede that these words may resonate with the force of their other possible past and future meanings or usages. The author, for those who have, willingly or unwillingly, been co-opted into the intentionalist camp, has persisted as a means of curtailing and controlling the unruly rhetoricity and plurivocity of the text.

Yet this is dramatically at odds with the wholesale critique, not only of the author, but of the modernist subject itself (of which the author could be said to be a localized manifestation), in the late twentieth century. Often, this critique of subjectivity has included within its remit a critique of intentionalism and the intentionalist project. My contention here is that these two – intention and subjectivity – need not be so facilely run together. What remains to be discussed is whether we might conceive of some intention *after* the subject, and what form this 'intention of the text' or 'intentionality of form' might take. The transition, which I am attempting to plot, entails the re-conceptualization of intention as material, linguistic, textual (of the text) rather than mental, subjective (of the subject). In turn, this will open up the possibility of an 'ethics' of the text, in contrast to the limited anthropocentric ethics of Hirsch and his latter-day intentionalist imitators. At the close of the first chapter I expressed a desire to move from a consideration of authorship to one of agency and – more radically – towards a vision of agency (and ultimately intention, conceived as an animating principle or purpose, or purposiveness, of the text) as a formal rather than a human or subjective quality or state. The structuralist and post-structuralist critiques of the subject, I will suggest, both facilitate (perhaps unwittingly) and necessitate this manoeuvre. This is despite the fact that on the face of it they appear to be unequivocally anti-intentionalist; such a judgement requires us to specify what kind of intentionalism it is that they are opposing, what version of intention it is that they are abandoning – my answer is that it is a subject-centred one. Furthermore, I would argue that it is possible to plot a trajectory from Wimsatt and Beardsley's limited anti-authorialism to this more recent critique of subjectivity: the latter is merely a stronger and more generalized version of the former, and like its forebear it falls short of ruling out intention per se.

But first, let us consider what this much vaunted disappearance of the subject entails. More properly and pertinently, what has taken place is an interrogation of the singularity and authority of the subject – those features of it, in fact, considered requisite for authorial intentionalism. Any admission of flux into the being and boundaries and location of the subject will trouble its ability to fix the meaning – and value, perhaps – of the text from which it has, in any case and of necessity, been sundered.[1] In *Discerning the Subject*, Paul Smith offers the following definition:

> The 'subject' is generally construed epistemologically as the counterpart to the phenomenal object and is commonly described as the sum of sensations, or the 'consciousness', against which the external world can be posited [. . .].

> In different versions the 'subject' enters a dialectic with that world as either its product or its source, or both. (1988: xxvii)

With the dwindling of epistemological certainties have come challenges to the rationalist epistemological paradigm of the subject–object relation. Specifically, what has been challenged is 'the role of the "subject" as the intending and knowing manipulator of the object, or as the conscious and coherent originator of meanings and actions' (xxviii). This is exactly the role that the Hirschian and Searlian subjects are expected to play in their production and delimitation of meaningful texts and utterances. Smith attributes this unsettling of the subject to 'poststructuralism' (the quote marks are his), but in fact the sceptical rot set in before the publication of anything we might reasonably term 'poststructuralist', and we can trace the anti-subjectivist trend in criticism back through structuralism, New Criticism, Russian Formalism and the work of Saussure.

The already compelling case against the subject is consolidated by Lacanian structural psychoanalysis, with its emphasis on the linguistic nature of the unconscious and the role of this in 'producing' the subject. Jean-Marie Benoist claims that 'both the notion of man as a specific *object* of knowledge, and as *subjectivity* – the clear and lucid source of all the meanings arrayed around him – are challenged at their very foundations' by the Lacanian theory of the (split) subject, and the determinative role of the unconscious means that the object of knowledge is now:

> the unconscious as a bundle of deep syntactic patterns; it is the deep structures of kinship and the systems of transformation which generate such and such a classification of concrete data within a totemic system; it is the relationship between deep structures and surface structures in linguistics. Man, in other words, is the great absence. (1978: 18)

He is 'absent' because he has been usurped by structures and systems over which man – a recent historical invention, a function of a particular post-Enlightenment way of interpreting the world – can exercise no control. 'Man' is constructed within and by language and semiology more generally: is a sign within the system and cannot exist independently of this system – at least not as 'man'.

Seán Burke, however, expresses alarm at the Lacanian suggestion that 'it is not man as conscious subject who thinks, acts or speaks, but the linguistic unconscious that determines his every thought, action and utterance', and he questions the apparent seamlessness of the passage from one anti-subjectivist formalism to another in the twentieth century, citing as a significant interruption of this logic, the intervention and influence of phenomenology in the 1940s and 1950s which sought to re-establish this conscious subject as the 'ground of knowledge' and the source of perception (1998: 13, 10). This point is borne out by my analysis of Iser and Poulet in the preceding chapter, but that analysis revealed also the potential for a reconfiguring of the concept of intentionality which might distance it from its subjectivist associations.

Whether or not Burke is correct to dispute the perceived linearity and coherence of anti-subjectivism, it is certainly true that what began, in each case (New Criticism, Structuralism, etc.), as strategic statements of linguistic and textual autonomy – an autonomy from the sphere and influence of the subject and the attendant baggage of biography and expressivism, for the most part methodologically motivated – have gradually come to be interpreted as a sustained and progressive attack on subjectivity itself or, at the very least, a decentring and destabilizing of that subject. Anti-subjectivism has only retrospectively cohered into a 'movement'. Linda Hutcheon, for one, protests that postmodernism, rather than being the nail in the coffin of post-Enlightenment subjectivity, instead serves to 'situate' and 'historicize' the subject – 'decentering' it certainly, and 'pointing directly to its dramatized contradictions', but not actually 'denying' it (1988: 159). The talk of 'death' and 'disappearance' – the virulence of the 'negative theology' which Burke takes issue with (1998: 26–27) – may be overblown, but it indicates the magnitude of what literary critics (and indeed all interpreters of action and utterance) are being asked to do without: not only practical criteria for fixing those interpretations, but also that instinctual mapping of consciousness onto language and/or texts (which Poulet does, most obviously) and the tendency to read those texts in the light of some posited consciousness, real or imaginary. Can these cognitive and agential elements be dispensed with altogether, or do we then begin to attribute a compensatory kind of subjectivity to the structures and systems of which Benoist speaks? This is a question which any theory of structural intentionality will have to answer, in shifting the locus of intention from subject to text. What, moreover, do we mean by 'cognitive' in the context of a text that has been sundered from the writing subject and from that subject's history? Paul de Man's claim that 'it follows from the rhetorical nature of literary language that the cognitive function resides in the language and not the subject' (1983: 137) will be examined later in this chapter: this, to my mind, initiates a move towards a theory of conscious agency located within and emanating from the text itself but one which, unlike Poulet's mapping of subjective consciousness onto the text, takes account of its form.

The metaphysics of presence: The presence of intention

But first it is instructive to consider how one theorist of the late twentieth century deals with writing which is no longer, according to Foucault, 'the insertion of a subject into language' but rather serves to elide and override that subject (1977: 116). For Jacques Derrida, certain versions of the subject are unequivocally dispensable, notably that classical, omnipotent subject who is seen to produce and fix meaning:

> The 'subject' of writing does not exist if we mean by that some sovereign solitude of the author. The subject of writing is a *system* of relations between strata: the Mystic Pad, the psyche, society, the world. Within that scene, on that stage, the punctual simplicity of the classical subject is not to be found. (1978: 226–27)

In attacking the 'punctual simplicity' of this 'sovereign' subject/author, Derrida reveals that what is problematic for him about such a subject is its immediacy, its determinacy, its simple *presence* above and beyond that 'system of relations between strata' of which he speaks. Throughout his work, even when not seeming to address the issue of subjectivity, Derrida implies that this Western philosophical desire or nostalgia for the subject is symptomatic of a wider hankering after *presence*, and it is against the 'metaphysics of presence' that his influential early philosophy is directed (1973: 51). This has a direct bearing upon the issue of literary intention, because the analysis of the concept in the preceding chapters has suggested that intentionalists and anti-intentionalists alike have erred in figuring intention as presence, specifically: the presence of the subject in the text (or not, as the case may be), the 'voice' of the author, his blueprint intention made manifest (*present*ed), the immaterial materializing or force taking form, mind becoming matter; the pitfalls and agonies of this transition or metamorphosis (this *making present*) have been extensively discussed. What else does E. D. Hirsch mean by intention being 'the soul of speech' (1976: 90)? Such an expression represents intentionalist critics as engaged in something akin to a spiritual quest, desirous of hearing the pronouncements of 'the oracle' (Wimsatt and Beardsley 1962: 104), deliverer of what will be, in critical or interpretative terms, the last word.

Even those not directly concerned with making a case for intentionalism have operated within the metaphysics of presence or, more pertinently, according to a law of present (to mind) intention. One of the pre-requisites of a successful performative, according to Derrida's reading of Austin is:

> the conscious presence of the intention of the speaking subject for the totality of his locutory act. Thereby, performative communication once more becomes the communication of an intentional meaning, even if this meaning has no referent in the form of a prior or exterior thing or state of things. This conscious presence of the speakers or receivers who participate in the effecting of a performative, their conscious and intentional presence in the totality of the operation, implies teleologically that no *remainder* escapes the present totalization. (1982: 322)

As detailed in Chapter 2 here – and setting aside for a moment the viability of Derrida's reading of Austin – the speech act theorist customarily comes all too close to asserting the presence, and self-presence, of the correct intention as the primary criterion for a successful performative, regardless of other circumstances. As we saw, the impossibility of establishing this presence-to-mind of the necessary intention, which is really the impossibility of deciding absolutely whether the speaker 'means' what he says (for a lie may be outwardly indistinguishable from a genuine promise), causes problems for Austin. Such potential infelicities and non-standard speech situations ultimately threaten the supposed standard or ideal speech situation from which they deviate. Rather than excluding such deviations from his analysis of speech and meaning, Derrida shows himself to be above all fascinated by this 'remainder' which escapes 'the present totalization'. Everything that escapes the totality throws into doubt its 'total' nature and, therefore, the possibility of pure presence.

It is in *Speech & Phenomena* that Derrida first and most forcefully contests the traditional interpretation of being as presence which he judges as central not only to Western metaphysics but also to the work of those, such as Husserl, who purport to operate outside traditional 'metaphysical' philosophy. Thus he sets out to uncover and critique Husserl's (and phenomenology's) commitment to 'the original self-giving evidence, the *present* or *presence* of sense to a full and primordial intuition' as 'the source and guarantee of all value' (1973: 5). Phenomenology (and the Western philosophical tradition generally, in Derrida's opinion) asserts as the ground of all knowledge, the self-knowing, self-present subject; being *just is* being present (to oneself). Traditional intentionalism similarly asserts the self-presence and transparency to self of the originating author; it is conscious rather than unconscious intentions which count and the work is seen as emerging out of this state or stage of self-consciousness when the author wills the work into being. Husserlian phenomenology is shown by Derrida to be dependent upon this understanding of being as presence: with its privileging of apodictic evidence, which is the 'presentation [. . .] of objects to an immediate and self-present intuition' (Allison 1973: xxxiii) such that no further justificatory evidence or argument is required; but also with its idea of transcendental consciousness, which is neither more nor less than 'the immediate self-presence of this waking life' and which exteriorizes all that is 'empirical, worldly, corporeal', everything which is somehow secondary or mediated (ibid.). Husserl's distinction between expression (*Ausdruck*) and indication (*Anzeichen*) also invokes this motif of presence and this logic of inside and outside in its categorization of different levels of meaning: while *Ausdruck* stands for meaning in a pure, self-present state, *Anzeichen* is more like reference or applied meaning.

This expression/indication distinction is one which Hirsch arguably borrows and adapts for his own meaning/significance distinction in *Validity in Interpretation* and which has echoes in other intentionalist theories where authorial meaning takes the form of an intentional blueprint existing prior to any material signification and has a purity which will not, so the argument goes, be corrupted by that process of making meaning manifest and legible; it may be added to – as the text gains additional significance(s) over time and for different readers – but the pure expressive kernel of authorial intended meaning remains inviolable, uncorrupted. In communication, Husserl declares, there is an 'interweaving' of expression and indication, the argument being that the latter exteriorizes or *re-presents* 'what primordially occurs in [the] inner sphere' of expression (Allison 1973: xxxv). So expression is thereby figured as a solitary mental act – rather as the text, as a material, publicly available artefact, re-presents that initial, private intentional blueprint of the author, according to intentionalist critics in search of determinate meaning. But despite this interconnectedness of expression and indication in communication, there is, according to Derrida's reading of Husserl, a higher value placed on the primordial or ideal meaning that is 'expression' than on the meaning (indication) which escapes this solitary sphere of experience. Husserl attempts to isolate this realm of expression as something purer than and prior to communication: thus interior monologue is held up as the paradigm case of pure expressivity, stripped as it is of the 'non-expressive features of language (such as its physical aspect and its communicative potential)' which are judged merely 'contingent accidents' (Howells 1999: 20); and thus silence and solipsism are heralded as a kind of perfection. Such interior expression is also 'present' in a temporal sense, that is, in its instantaneousness: it is both immediate and outside time. Because of the mapping of being onto presence, absence (non-presence) is automatically associated with the negativity of non-being: death or, in the case of texts and utterances, meaninglessness.

Despite this phenomenological accent on presence and ideality – which enables the repetition of meaning, time and again, in uncorrupted form, the repetition of the 'same' – Derrida observes that phenomenology is, in fact:

> tormented, if not contested from within, by its own descriptions of the movement of temporalization and of the constitution of intersubjectivity. At the heart of what ties together these two decisive moments of description we recognize an irreducible nonpresence as having a constituting value, and with it a nonlife, a nonpresence or nonself-belonging of the living present, an ineradicable nonprimordiality. (1973: 6–7)

What is impossible, he insists, is to isolate this moment of pure presence, in its singularity and immediacy, for it is always already part of some (communicative) system of relations. So expression, for Derrida, is always already tainted by indication, by its communicative and significatory potential – which potential is a necessity (this idea of necessary possibility is a recurring theme in Derrida's writing). Husserl's 'interweaving' becomes, in Derridean terminology, an 'entanglement' (20): while the former term suggests both the separability of the 'threads' which are being woven together and the orderliness of their combination, the latter implies instead a messiness and difficulty of disentanglement. Derrida posits that 'the discursive sign, and consequently the meaning, is *always* involved, always *caught up* in an indicative system. Caught up is the same as contaminated' (ibid.). So presence is always interrupted by absence and there is no pure, originary, interior, self-present, pre-communicative meaning which we might separate off and analyse independently of its public, signifying function.

Although Husserl means something very particular by 'expression', yet it is evident that some notion of self-present meaning is required for the success of traditional authorial intentionalism. The blueprint conception of intention is of something complete and impermeable which exists perhaps even before the author has committed a single word to paper and which then is re-presented in material form: the text mimics the interior (mental) presentation of meaning. What critics of intentionalism have challenged is both the propriety of this act of re-presentation or mimicry – citing the complications involved in making public what is private – but also the self-presence of this originating intention, and its transparency to the author doing the intending. Derrida's attack on the phenomenological doctrine of expression would appear to back them up in this.

It is the desire for presence, Derrida argues, which has led to the overwhelming emphasis on phonetic writing within Western philosophical culture. Central to this is the belief in the, 'system of "hearing (understanding)-oneself-speak" through the phonic substance – which *presents itself* as the nonexterior, nonmundane, therefore nonempirical or noncontingent signifier' (1976: 7–8). What this self-presence of the voice ensures is the immediate transmission of knowledge, certain knowledge of the self or subject and, crucially, certain knowledge of the intentional agency which characterizes and expresses this subject. Writing, then, is perceived as merely the material representation of the more authentic, instantaneous, original voice; it has been confined, traditionally:

> to a secondary and instrumental function: translator of a full speech that was fully *present* (present to itself, to its signified, to the other, to the very condition of the theme of presence in general), technics in the service of language, *spokesman*, interpreter of an originary speech itself shielded from interpretation. (8)

Speech is 'shielded from interpretation' because the voice is, so the argument goes, direct, present, unmediated, not in need of interpretation in the way that writing – which is premised on the absence, potential or actual of the author and reader – is. In addition, writing is seen as contrived or artful: technique rather than authentic expression. The disjunction between speech and writing, as we saw in Chapter 2, made the application of speech act theory to literature more troublesome than it might have been, precisely because the connection between speaker and word is apparently broken and the conventions of face-to-face conversation are violated. It is interpretation, that perilous enterprise, which causes the transmission of knowledge (of the subject) to falter and – according to phonocentric lore – writing is more susceptible to interpretation (and re-interpretation) than speech. Writing, therefore, threatens a corruption of the purity and self-presence of meaning and this 'pure' meaning exists in its utmost purity in the form of the speaker/subject's intention-to-mean, which exists ideally in the form of the unheard interior monologue. Poulet's solution to this dilemma, as previously delineated, was to allow the interior monologue to be heard, via a sharing or meeting of consciousnesses, the transcendent nature of which belied its dependence upon some publicly sharable, substantial set of signifiers: the book.

Derrida describes this phonocentric philosophy as having its roots in Aristotle's pronouncement that 'those that are in vocal sound are signs of passions in the soul, and those that are written are signs of those in vocal sound' (Aristotle 1962: 23). So, the voice represents what is in the soul – a more modern translation might substitute 'mental experience' for 'soul', as Derrida does (1976: 11) – while writing merely represents the voice. Too often, intentionalist criticism has involved an attempt to recapitulate the mental experience of the author in creating the text – Hirsch's term is 're-cognise' – thus tapping into some supposedly pre-significatory understanding between the author and his ideal or intended reader. For these 'passions of the soul' are universal – 'the same for all' says Aristotle (1962: 23) – even though the vocal and written signs which represent them are not; the passions are natural, the signs conventional. In this way, a distinct hierarchy is erected (nature good, convention/culture bad or distorting), and we take on the idea that the voice, despite being a signifier of sorts, 'has a relationship of essential and immediate proximity with the mind' (1976: 11) – or the soul, if you prefer – that is, a natural, non- or pre-conventional relationship.

For my purposes, in evaluating the history of intentionalism, the voice is traditionally viewed as the primary and most proximate signifier of the meaning-intention and, therefore, an expression of being itself. Why? Because being manifests itself as presence and knows itself as such through the self-presence of consciousness, and because intentionality – the directedness of consciousness upon its objects – is the mark of the mental. So close is the voice to what it signifies that it practically effaces itself (this is the blessed immateriality of the voice, its transcendent and quasi-spiritual character); it is the vehicle for the communication of the subjective intention and, as such, is ideally transparent: '*Phonē* [. . .] is the signifying substance *given to consciousness* as that which is most intimately tied to the thought of the signified concept. From this point of view, the voice is consciousness itself' (Derrida 1981b: 22). When Husserl claims that 'the voice simulates the conservation of presence' (Derrida 1973: 15), he means by 'voice' something particular to phenomenology (although influential on subsequent phonocentric philosophy):

> It is not in the sonorous substance or in the physical voice, in the body of speech in the world, that [Husserl] will recognize an original affinity with the logos in general, but in the voice phenomenologically taken, speech in its transcendental flesh. (16)

An interesting oxymoron, this, suggesting something neither fully material (flesh) nor fully immaterial (transcendental):

> the intentional animation that transforms the body of the word into flesh, makes of the *Körper* a *Leib*, a *geistige Leiblichkeit*. The phenomenological voice would be this spiritual flesh that continues to speak and be present to itself – *to hear itself* – in the absence of the world. (ibid.)

So, what is suppressed in Husserl's account of the phenomenological voice as 'spiritual flesh' is the material or substantial aspect of the physical voice, in favour of the immateriality of an interior mental voice; it is 'intentional animation' which turns body into spirit (a kind of de-substantiation).

But this idea of 'voice' as self-present meaning figures also in intentionalist literary criticism, where, audio books and public readings aside, there is no question of the actual, physical voice of the author being heard. It suggests the expressive character of the text: the author may 'voice' his ideas through the narrator or through a particular character, or his 'voice' may be 'heard' most clearly at a specific point in the narrative (this 'hearing', to the extent that it can be located at all, occurs within the consciousness of the reader, so both speaking and hearing are rendered immaterial and, we might say, metaphorical). This attribution of a 'voice' to the text implies also a kind of unity, subjective and stylistic: the text as the expression of a unified speaking subject whose distinctive 'voice' is recognizable in other works, and can be distinguished from that of other authors. The persistence of the metaphor of voice is remarkable. Even if replacing the 'author' with some more strategic and sceptical 'author-function', the tendency is still to see the text as imbued with a single and singular voice: transcendent, omnipotent, self-aware. Even where a work is plainly contradictory and plurivocal, these different voices will be figured as having a single origin – the very idea of origin suggesting something outside the text itself. Yet the text does not so much 'speak' the author's meaning as ventriloquize the critic's in the author's absence (giving rise to notions such as the 'hypothetical' author, the waxwork complement to the ideal reader).

This physical absence of the author is an absence which Derrida claims 'belongs to the structure of all writing' and 'all language in general': 'The absence of the sender, the addressor, from the marks that he abandons, which are cut off from him and continue to produce effects beyond his presence and beyond the present actuality of his meaning' (1982: 313). Interestingly, the Samuel Weber translation of 'Signature Event Context' in *Limited Inc* has the author cut off from 'the present actuality of his *intentions*' (1988: 5; my emphasis) and the equivocal translation of *vouloir-dire* throughout Derrida's works in English as either 'meaning' or 'intention' reinforces the inextricability of those concepts. However, to be cut off from the author's meaning/intention does not mean to be cut off from meaning (and, by extension, intention) altogether; the text does not therefore become meaningless – far from it. Rather than lament this, as we will see, Derrida celebrates the schism between (alleged) authorial self-presence and the material text, because he does not see it as dampening the force of writing or hindering it in its production of meaning(s), but rather as liberating the text from its attachment to a controlling subject and to a particular, originating, intention. What I want to suggest is that such a liberation does not therefore strip the text of its intentionality, in a sense derived from the phenomenological use of that term.

Writing: The mark as machine

To Husserl's assertion that 'Everything that escapes the pure spiritual intention, the pure animation by *Geist*, that is, the will, is excluded from meaning (*bedeuten*) and thus from expression', Derrida responds by claiming that 'the opposition between body and soul is not only at the centre of this doctrine of signification, it is confirmed by it' (1973: 35). Consequently, the written text – suggestive of corporeality and even a kind of carnality – is relegated to a position of secondariness within this system, as dead matter, inert, requiring an intention to animate it (just as an object requires some subject to intend it, to present it to consciousness) for meaning is mental (a question of spirit, or *Geist*) and writing is merely matter. Alternatively, writing is viewed as merely a 'vehicle' of communication, existing only to transport and display some 'idea' or 'signified content' (see Derrida 1982: 311ff.; and 1988: 3ff.) – or, as we saw in the account of reading given by Poulet, as brute matter to be transcended, turned into 'a series of words, of images, of ideas' existing in the mind of the reader (1969: 54).

Again, this serves to explicate both the desire to bring together intended and instantiated meaning that is evident in traditional intentionalism, and the difficulties implicit in this (such as the inaccessibility of the idea in its pre-instantiated form). These difficulties can now be viewed as the fallout of Western philosophy's radical disjunction of mind and matter (and, relatedly, force and form, subject and object, intention and text); the stubborn traces of Descartes' 'real distinction'. Nevertheless, in Derrida's re-thinking of the concept of text (and form and matter and so on), it begins to be possible both to account for the materiality of the text and to hold onto some notion of intentional agency at work in that text. For he asserts that:

> To write is to produce a mark that will constitute a kind of *machine* that is in turn *productive*, that my future disappearance in principle will not prevent from functioning and from yielding, and yielding itself to, reading and rewriting. (1982: 316; my emphasis)

What the mark produces, ineluctably, and beyond the interference of the author, is meaning: in fact a potentially limitless array of possible meanings, readings and interpretations. And by his 'future disappearance' as author of this mark he means, 'for example, the nonpresence of my meaning, of my intention-to-signify, of my wanting-to-communicate-this, from the emission or production of the mark' (ibid.). It is the intention-to-signify and wanting-to-communicate-this (in its this-ness) that intentionalist critics have sought in vain to reveal as present and visible in the text, but Derrida sees the very nature of writing as premised upon the *non-presence* in the text of these specifically authorial intentions, the *inaudibility* of the author's voice. This conception of text as an autonomous, generative 'machine' is more than Wimsatt and Beardsley could have bargained for with their tentative remarks about 'effective intention' or intention in the text and their endorsement of semantic autonomy. This is, rather, an intention *of* the text – because 'text' in the New Critical and post-structuralist uses of it, has quite different connotations: closed or complete on the one hand (the poem as a 'verbal icon', static in its iconicity), radically open on the other. Paradoxically, with their ruminations on the agency of the text itself – which, remember, is portrayed as an intentional, active subject in its own right – it is Poulet and his phenomenological ilk who have edged us towards this conclusion, although it cannot be said that they ever anticipated this usurpation of sovereign subjectivity by textuality.

In light of this Derridean view of the text as generative or productive, can the mark itself, then, be said to display a form either of intention – if this is seen as inextricable from meaning and the production of meaning – or of intentionality, in the sense that it is 'intent' upon (continued and varied) meaningfulness, in some manner? To begin with, we might question what it is for the mark to be a 'machine'. This is an interesting, de-humanizing metaphor, and suggests a kind of agency and dynamism of the text itself, a generative quality, a formal or structural rather than human or subjective agency. Intentionalism has always struggled with the apparent inaccessibility of authorial consciousness, but an agency which is mechanical will have nothing to do with consciousness – despite the origins of the machine in some human creative endeavour, the very term 'mechanization' implies the ability to function without the effort or interference of some person or other. What is not clear is to what extent – and by what or whom – this mechanical agency is contained or circumscribed. I would suggest that something or someone must limit – if only for the moment of reading and comprehending – the text's play of meanings, if Gasché's version of Derrida (as positing some kind of in principle determinable system, rather than a semantic free for all) is to win out (Gasché 1987).

The machine metaphor is used to powerful effect by Peter Brooks in his description of the dynamism of plot in *Reading for the Plot.* According to Brooks' post-structuralist narratology, the agency of the author is largely superseded by an agency latent in narrative itself, an agency displayed in the force of the plot:

> Plot as it interests me is not a matter of typology or of fixed structures, but rather a *structuring operation* peculiar to those messages that are developed through temporal succession, the *instrumental logic* of a specific mode of human understanding. Plot, let us say in preliminary definition, is the logic and dynamic of narrative, and narrative itself a form of understanding and explanation. (1984: 10; my emphasis)

So plots are 'intentional structures' rather than merely 'organizing structures' (12). This is suggested by the agency afforded them by the description 'structuring operation'; they are active rather than static, they display the unfolding of a logic of temporality.

The notion of an 'instrumental logic' at work is also reminiscent of a phenomenological conception of intentionality – the narrative tends towards, is intent upon, some object or end. It is the implied finiteness of this end that Brooks goes on to interrogate, while retaining the idea that the narrative displays a kind of forward momentum, which he characterizes as 'desire', thus:

> In [the] absence [of ends], or their permanent deferral, one is condemned to playing: to concocting endgames, playing in anticipation of a terminal structuring moment of revelation that never comes, creating the space of an as-if, a fiction of finality. (313)

And this (post-structuralist/post-modern) suspicion of and subsequent resistance to ends has had an effect upon the nature of plot, 'which no longer wishes to be seen as end-determined, moving towards full predication of the narrative sentence, claiming a final plenitude of meaning' (314). So the movement implicit in plots is not a movement towards 'plenitude' or the full presence of meaning, but merely a movement towards the next endgame or 'fiction of finality'; there is a self-consciousness about this process, an awareness that meaning is never completed, never final, that what ends there are, are pretended, chosen, random. Brooks is treading a delicate line here, between his intuition that 'the interminable would be the meaningless', that is, that ends have a necessary role to play in structuring the meaning of the whole (by making it a whole and allowing it to be distinguished from what it is not, from what comes before or after or beyond), and his post-structuralist leanings, which invoke a mistrust of finality and totalization (93). From this balancing act we can begin to develop an idea of intentional (directed) narrative momentum which is more or less teleological, while resisting claims to an objective or full meaning indicative of the kind of presence that Derrida has set out to criticize.

In the meantime, it is instructive to pursue the possible parallels between Derrida and Brooks further. By way of analysing the peculiar momentum of plots, Brooks considers the role of the motor in a number of nineteenth-century novels and in doing so offers a distinction between 'machine' and 'motor' which could prove useful in adapting and developing Derrida's conception of the mark as a 'kind of machine'. According to Brooks, who cites Michel Serres' book on Zola as inspiration, the machine is 'a system for the transmission of forces outside itself' while the motor, more pertinently, is figured as

> containing its source of movement within itself, built on the three principles of difference (of temperature), reservoir (of fuel), and circulation. The self-contained motor, working through combustion – typically, the steam engine – also corresponds to the emerging conception of human desire. (41)

It is on the basis of this definition of the motor that he argues 'the dynamic of the modern narrative text is that of the steam engine' and that such motors and engines, as well occurring *within* the plot (e.g., as subject matter in Zola and elsewhere), are 'representations of the dynamics of the narrative text, connecting beginning and end across the middle and making of that middle – what we read *through* – a field of force' (44, 47). It is desire which is the 'motor' of narrative: the desire for an end which, when it comes, to the extent that it comes at all, will be merely a strategic and temporary closure. This desire, furthermore, is subject to interruptions and obfuscations – in any text there are moments of delay, displacement, deferral where the forward movement of the plot is hindered – this is the 'field of force', the uneven passage from beginning to end.

What is significant about this reading of narrative is the way in which Brooks is building upon Derrida's work on force to suggest that it is something implicit in the form of narrative. Brooks at various points cites Derrida's lament in 'Force & Signification' that structuralist criticism is 'separated from force' (Derrida 1978: 5), declaring that we would, 'do well to recognize the existence of textual force, and [. . .] we can use such a concept to move beyond the static models of much formalism, toward a dynamics of reading and writing' (1984: 47). My contention, however, is that an expanded and improved formalism would make way for the description of force as something that inheres in form. Brooks (like Derrida) is correct to attack old style formalisms for their limitations – in particular for their conception of texts as displaying a kind of fixed and immovable structure – but Brooks himself is working within what was originally a formalist discipline (narratology) and attempting to recreate and refresh that discipline via an encounter with post-structuralism and psychoanalysis.

Unfortunately, for my purposes at least, Brooks ultimately shies away from attributing intentional agency to the text itself, and instead conceives of it as *relational*, as to do with the relationship between reader and text. Thus plot – 'the organizing line and intention of narrative' – is, he goes on to say:

> best conceived as an activity, a structuring operation elicited in the reader trying to make sense of those meanings that develop only through textual and temporal succession. Plot in this view belongs to the reader's 'competence', and in his 'performance' – the reading of narrative – it animates the sense-making process. (37)

The narrative opens certain questions at its outset, and so arouses a 'desire' for resolution in the reader which the subsequent relation of events and actions will seek to satisfy. Narrative desire in the modern novel can also be figured as that force which 'drives the protagonist forward', a kind of 'ambition', so there are different levels and manifestations of desire at work here (39).

Brooks' focus on the reader suggests that the intentional narrative momentum of which he is writing – this movement forwards, towards some nominal end – is something essentially relational. So then is the theory of formal intentionality that I am trying to elucidate something that inheres in this relationship between reader and text, rather than something specific to the text itself? Tentatively I would like to resist this conclusion, if only because it necessitates the presence of some intending subject for the production of meaning, and it appears to be a reader-response version of Knapp and Michaels' dismissal of the possibility of intentionless meaning (which is really a dismissal of the possibility of meaning that is severed from the subject). My assertion – against the odds, it seems – that there is a detectable intentional drive in the structure and formal make-up of texts is, by contrast, a more straightforwardly formalist conclusion, but harder to describe, and perhaps possessed of fewer significant ramifications as far as critical methodologies in literary studies are concerned. However, I am not attempting to delineate a new way of reading and interpreting texts, but rather proposing that it is in the nature of texts to be intentional and (therefore) that we read them as such (to do so is not a choice, it is merely what we do). So rather than develop some kind of critical methodology, I simply want to create an aperture for intention within our understanding of literary texts – and within the theoretical discourse around literature and literariness (or even to suggest that some notion of intentionality is already to be found in certain theories of literary meaning); which is not to deny that the relationship between text and reader may also be described as having intentional aspects.

Another potential problem with Brooks' approach is the mentalism of it, the importance of the consciousness of the author and reader. The text becomes for him in this way the site of an exchange or transaction of meanings – he notes the 'contractuality of narrative' – and it is this which facilitates the development of the transference analogy that Brooks falls back upon (225, 216ff.). My appeal to Derrida's conception of the mark as 'machine' represents a concerted attempt to move away from the model of (human) consciousness, from accounts of the text as either a form of subjective consciousness in itself or as the meeting place of the consciousness of author and reader, respectively. The reasons for such a move should, by now, have become apparent: not least the ease with which the mental becomes mystified and mythologized (and so rendered either inaccessible or somehow untheorizable) and the undesirability of anthropomorphic accounts of literary form in a post-structuralist climate.

Fundamentally, though, it is the focus on plot which means that Brooks' idea of desire in narrative does not quite coincide with my conception of textual intention. Although Brooks does not identify 'plot' with the content or themes of the narrative, it has enough of those connotations to move us a step away from the desired emphasis on language and linguistic or formal structures; his avowed anti-formalism is, perhaps, to blame here. Is it only 'narratives' – texts, verbal or written, which can be said to have some kind of 'plot' or storyline – that display intentionality, or will any collection of written words do this? An ancillary concern here is that, although Brooks is primarily discussing written texts, the term 'narrative' nowadays implies a whole range of different possible media and forms which could feasibly submit themselves to a narratological analysis, whereas I am keen to home in on how *texts* work, rather than, say, films or comics or stained glass windows or paintings – to borrow a few examples of narrative from Barthes' seminal essay on the subject (1977: 79).

What is most useful about Brooks' take on narratology, however, is the insistence upon a reconceptualization of the relationship between force and form which recognizes their fundamental inextricability, and this aspect of his thesis he derives from Derrida. Poulet, as we have seen, conceives of the text as 'a sort of human being [. . .] a mind conscious of itself and constituting itself in me as the subject of its own objects' (1969: 59); but Derrida's text, remember, is a 'machine' rather than a human being. Although both Poulet and Derrida figure the text as possessed of an intentional agency (an ability to 'intend' objects and meanings, in the phenomenological sense of intentionality), the latter attempts to move away from the subjective slant of phenomenological criticism and from the view of the text as something ideal and immaterial. For Poulet's focus is on *force* to the exclusion of *form*. By contrast, Derrida, I would suggest, wants to have it both ways: in addition to his attack on so-called 'idealist' criticism he is (as Brooks noted) elsewhere notably critical of the structuralist suppression of 'force' lamenting the fact that 'Criticism henceforth knows itself separated from force, occasionally avenging itself on force by gravely and profoundly proving that separation is the condition of the work, and not only of the discourse on the work' (1978: 5). But separation, for Derrida, is manifestly *not* the condition of the work (which has no 'outside' that we could attempt to delineate), because such separateness implies, first, a view of force as prior intentional blueprint or posterior addition and, second, the supplementary character of writing (in the traditional, metaphysical, negative sense of that term). Derrida persistently challenges the idea that:

> Reading and writing, the production or interpretation of signs, the text in general as fabric of signs, allow themselves to be confined within secondariness. They are preceded by a truth, or a meaning already constituted by and within the element of the logos. (1976: 14)

It is this idea of origin – and originating truth, the first (and last, and unchallengeable) word – which Derrida unsettles; writing becomes something which is without origin, which has always already begun, and which is, furthermore, *everywhere*. So, in *Of Grammatology*, Derrida famously testifies that:

> *There is nothing outside of the text* [there is no outside-text; *il n'y a pas de hors-texte*] [. . .] there has never been anything but writing; there have never been anything but supplements, substitutive significations which could only come forth in a chain of differential references, the 'real' supervening, and being added only while taking on meaning from a trace and from an invocation of the supplement, etc. And thus to infinity. (158–59)

So writing, rather than finding its origin in the subject, produces that subject, which must henceforth be viewed as part of the text, within archi-writing rather than outside it; it is in this way that the subject is 'situated' rather than 'destroyed' by Derrida, as previously cited (Macksey and Donato 1970: 271). Derrida's reading of Rousseau in *Of Grammatology*, then, 'must be intrinsic and remain within the text' rather than looking to some 'psychobiographical signified, or even toward a general psychological structure that could rightly be separated from the signifier' (1976: 159) – in fact it is not clear how it could do otherwise, given the recently propounded ubiquity of such signifiers – but what it is to be 'intrinsic' or to remain 'within' the text must be understood here in relation to this claim of the impossibility of an 'outside-text'; the boundaries of 'form' have shifted.

Thus, even the 'general psychological structure' is produced by language, and he goes further in arguing that there are all and only signifiers. Where this leaves intention is within the system of signifiers, but not dominating it, as previously thought. So, according to his proposed 'differential typology of forms of iter-ation': 'The category of intention will not disappear; it will have its place, but from this place it will no longer be able to govern the entire scene and the entire system of utterances' (1982: 326). And because of the necessary possibility of re-iteration ('iterability'), that is, of the repetition of the same words in different (unexpected, even unimaginable) contexts and in different ways, giving rise to different and even contradictory meanings: 'The intention which animates utterance will never be completely present in itself and its content. The iteration which structures it a priori introduces an essential dehiscence and demarcation' (ibid.). So, rather than the utterance being produced and structured by some 'completely present' intention, it is the essentially iterable nature of utterances which produces/structures the meaning-intention such that it can never be fully present, in the sense of intending a singular and complete meaning.

'Text', according to this understanding of it – as inclusive and expansive, always spilling over its own borders – is therefore the meeting place of form and force; more importantly, this 'force' is something which inheres in form, and so cannot simply be identified with either authorial or reader's intention. Rather than reading this as indicative of a terminal anti-intentionalism on Derrida's part, I would like to identify it as the very place where we might want to bring intention back into the account of literary meaning, as an agency of the letter itself, in its intentness-upon meaning. For it is writing itself, not some mental state of the author, which Derrida describes as 'inaugural', and he says, 'no knowledge can keep it from the essential precipitation toward the meaning that it constitutes and that is, primarily, its future' (1978: 11). This 'precipitation' towards an unknown and undeterminable future meaning I am interpreting, therefore, as a kind of radicalized intentionality of the phenomenological kind (a being-intent-upon) – radicalized, because of the posited indeterminacy of its outcome (the meaning that is its 'object' is never finally settled), and because of its attention to the form of the text, the mark or gram.

Unlike traditional intentionalism, this budding deconstructive intentionality does not provide us with some fully worked-out critical methodology, indeed its emphasis on the temporal open-endedness of meaning (intention as *movement* rather than *moment*, text as *productive* rather than *mimetic*) rather militates against this; in so doing it proposes an ethics of reading which is not in the humanist mould, but is rather a more general ethics of the open-ended, without being actually anarchic. But what it primarily indicates to me – and this is only the first step in a new way of thinking about literary form, of thinking beyond the critical paradigm of inside/outside – is that meaning is always somehow bound up with the idea of intention, and a reference to intention need not entail a reference to some murky and inaccessible subjective consciousness which is *outside* or which precedes the text.

Christopher Johnson observes that, in the 1960s, 'There is what might be called a minor epistemic shift within structuralism – or between structuralism and "post-structuralism" – with respect to the linguistic analogy, involving a change in emphasis from "language" to the more specific notion of "writing"' (1993: 4). Derrida is not alone, then, in attending as much to the textuality of structure ('structure' as metaphor) as to the structure of text – and Johnson is interested in particular in detailing the influence of information technology (such as cybernetics) upon Derrida and others. Thus, while structuralism tends to look to 'the "closed-system" models of linguistics and mathematics', which fit with 'the structuralist ideal of the homeostatic system', Derrida is more progressive in devising a conception of writing which owes more to 'the metamorphic and adaptational ("open-system") models found in systems theory', which are, presumably, use and context-dependent (7, 8). The 'adaptational' model, with its insistence on open-endedness, again points to the generative or productive nature of texts and their consequent, necessary incompleteness – a preoccupation with questions of closure and belonging and openness and structurality runs throughout his work, as has been amply demonstrated by Simon Critchley in his *Ethics of Deconstruction* (1992).

Traditional intentionalism, by contrast, is premised upon the finality of the text, its formal and semantic boundedness, which corresponds to the completeness of the intention which shaped it; in fact, one is isomorphic with the other, at least according to the most ascetic version of the blueprint theory. This clinging to closure is also metaphysical, in its insistence upon the propriety of the text or, in other words, in its absolute present-ness. But, to repeat this point, the meaning which Derridean texts are intent upon is never fixed (although it is not wilfully arbitrary either – its purposiveness perhaps precludes such an allegation) because this possibility of absolute presence has now receded in favour of a vision of textual becoming. Again, it is feasible to hark back here to the pronouncements of Poulet regarding the open-endedness of the literary work, whether considered as a form of subjective consciousness, or as a writing intention awaiting some temporary completion by/from its as yet unknown reader.

The mark's iterability also has ramifications for this vision of the text as open-ended process. Derrida had earlier used the emphasis on repetition and iterability to challenge the viability of Husserlian presence, arguing in *Speech & Phenomena* that 'the presence-of-the-present is derived from repetition' and not vice versa (1973: 52). The need for signs, written or otherwise, to be repeatable, necessitates *both* a kind of sameness and a kind of impurity which corrupts this sameness – the necessity of re-presentation haunts every sign such that every presentation is already a re-presentation; repetition introduces the difference which is a threat to presence. This is a threat unrecognized by Hirsch in his theory of type meanings, where the use and contextualization of the token is not seen as challenging the determinacy of the type which it re-presents in any way. Perhaps because he thinks of tokens as presenting (immediately instantiating), rather than re-presenting types, what Hirsch consistently overlooks is the disjunction between creative context and reading context; his responsibility to the author means that the former always overrides the latter.

Texts in particular are subject to this demand of re-presentation and re-reading. It is because of the knotty, public, material, persistent-through-time nature of texts that intentionalist critics have habitually latched onto the moment of creation (a 'now' of sublime self-presence for the intending author, in Husserlian terminology) as a fixed point or origin. This is a now which allegedly produces an ideal meaning which context, material alteration and readerly wilfulness cannot substantially change. And yet it has proved impossible to isolate this 'moment' of creation – the moment of intention realized (where and how?) – in its supposed singularity; such moments, because ideal, function like Husserl's silent interior monologue of expression and are not, therefore, accessible. Or, more accurately, they are only accessible in the re-presented, supposedly secondary form of the text itself. What is most problematical about texts, as we have seen, is what Blanchot, for example, has identified as their *incompleteness for the author*, and this sits uncomfortably with the traditional conception of intentions as complete or finite and determinative of something similarly finite. (Blanchot 1982)

Authorial intention, then, when viewed from the perspective of the doctrine of iterability and the ubiquity of writing, could be neither the source of textual meaning nor separable from it, because it will be itself, to some degree, textual or sensible, already part of the 'fabric' of signs, and always incomplete (or in the process of becoming). Above all, Derrida attacks the 'theological' conception of a meaning which precedes inscription, claiming that 'meaning must await being said or written in order to inhabit itself, and in order to become, by differing from itself, what it is: meaning' (1978: 11). In this way – as the epigraph to this chapter reveals – he moves away from the worship of some prior Idea or 'interior design' which a text exists to ex-press (that is: exteriorize). What this does not amount to, I have argued, is an abandonment of the concept of intention altogether, for we might say that inscription makes meaning and, as such, is possessed of a certain agency or intention. What this manner of thinking does endorse is a deconstruction of that tricky mind/matter distinction, where intention has always unequivocally been situated on one side of the equation; 'deconstruction', as should be increasingly evident, is not a reversal of the status of these categories, but a blurring of the boundaries between them, in this way accentuating their interdependence and inextricability. What Derrida leads us towards is an understanding of the materiality – even the textuality – of mind, and it is this which clears the way for an understanding of the textuality of intention and the intentionality of text.

Différance: Temporality and intentionality

We find throughout Derrida's work a subtle undermining, which is frequently a running together, of spatial and temporal models – in his 'fields' and 'epochs', for example, which are not simple spaces or periods of time. When, in 'Force and Signification', Derrida turns his attention to structuralism, he interrogates the mind/matter and force/form oppositions which I have described as foundational to literary criticism in the twentieth century and which have presented critics with the choice of one of two basic, mutually incompatible critical approaches: either formalist or psycho-biographical/intentionalist/sociological (take your pick). Traditionally the former have sidelined if not actually forbidden reference to intention, while the latter have made it integral to their understanding of texts. In particular, here, Derrida takes issue with the 'ultrastructuralism' of Rousset (and others) for granting 'an absolute privilege to spatial models, mathematical functions, lines, and forms' and for using 'structure' to refer 'only to space, geometric or morphological space, the order of forms and sites' (1978: 15). This structuralism which negates force in favour of form is the descendent of those anti-subjective formalisms which replaced biographical and intentionalist criticism, but in performing this act of negation they repress and contain whatever does not cohere with the 'geometrical model' in order to account for the work as a structural totality, a unified whole (17).

Derrida's interest lies precisely in what is repressed by this process of 'schematization and spatialization', namely a *force* which is not the product of some self-present subject and the *duration* of the work (its existence and changeability through time) (5). Thus force is figured as something which, contra structuralism, is instantiated or actualized in form and this instantiation occurs, not at the inception or inscription of form, the moment of creation, but continually, through time. My own suggestion is that it might benefit us to conceive of literary texts, analogously, as embodying or imbued with intentions which do not emanate either from some outside source or from some organizing centre (thus rejecting both the standard intentionalist/biographical and the structuralist position in favour of some alternative explanation). This involves seeing the work as purposive but not complete, not a totality, for – as regards duration, 'that which is pure qualitative heterogeneity within movement' (21) – Derrida declares that the literary work is never, as the structuralists aver, fully present, but always in the process of becoming. What critics should attend to is, then:

> the history of the meaning of the work itself, of its *operation.* This history of the work is not only its *past,* the eve or the sleep in which it precedes itself in its author's intentions, but is also the impossibility of its ever being *present,* of its ever being summarized by some absolute simultaneity or instantaneousness. This is why, as we will verify, there is no *space* of the work, if by space we mean *presence* and *synopsis.* (14)

And it is because of the structuralist literalism in thinking the notion of structure that this *temporal* element of the text, which precludes it ever being a quantifiable 'space', is largely overlooked in favour of the ideal of simultaneity which proffers the hope of a definitive, formalist interpretation. Against this doctrine of the now, Derrida argues the intransigence of the book, which is, 'first and foremost, volume' – by which he means: voluminous, extended, incomplete – the site of 'infinite implication, the indefinite referral of signifier to signifier' and so resistant to the imposition of this logic of simultaneity (25). Thus form fails to stifle force, regardless of the efforts of the structuralists; there will always be some remainder which resists totalization.

As an alternative to this static, spatial analysis of linguistic structures, Derrida gives us '*différance*' which is, famously, 'neither a word nor a concept' and which 'has neither existence nor essence' but which connotes nevertheless a movement or play of the difference and deferral implied by this handily crafted neologism (1982: 3, 6). Notably, its very nature as neologism is visible only in its written form, a cunning piece of Derridean impropriety – he himself calls it 'a lapse in the discipline and law which regulate writing and keep it seemly'– its inaudibility is really a blow against the phonocentrists (3). In delivering the '*Différance*' paper as a spoken address, Derrida must specify 'with an e' or 'with an a' with each use of difference/*différance* and so he is continually required to refer 'to a *written text* that keeps watch over my discourse' (4). Writing here *inaugurates* meaning, to some extent, because the meaning of *différance* emerges out of the visible (on the page) disparity between 'difference' and '*différance*'; this unsettles the usual privileging of idea over speech and speech over writing, as he and his audience, 'will be able neither to do without the passage through a written text, nor to avoid the order of the disorder produced within it' (ibid.).

It is the 'disorder', the 'lapses' of language, its potential for unseemly behaviour (which is really behaviour ungovernable by authorial/speaker intention, meanings which exceed or which are unanticipated by this intention) that Derrida particularly wants to document. More importantly, he wants to present them as constituting the very essence and nature of language. This stands in stark contrast to both Austin and Husserl's relegation of such lapses and errata to the margins of language, usually on the grounds either that they fall outside the intentional remit of the speaker, or that they are motivated by the 'wrong' type of intention (e.g., the intention of an actor on a stage to perform and entertain, rather than to convey some truth). Derrida, in fact, celebrates the unruly play of signifiers, a 'play' which, according to my reading of it, connotes an intentional agency written into the very structure of language, a generative potential of language. This 'play' does not, however, indicate an anarchy or absolute lack of system, for: '*Différance* is the systematic play of differences, of the traces of differences, of the *spacing* by means of which elements are related to each other' (1981b: 27). It is 'systematic' without, however, being regulated by some subject existing outside that system. This 'spacing', note, can be either spatial or temporal – although Derrida elsewhere unhelpfully avows that '*spacing* is neither space nor time' (1981b: 43) – and the emphasis on the spacing between elements of a system, rather than on those elements themselves, again underlines the centrality of absence to Derrida's philosophy: words mean, not on the basis of some inherent meaningfulness, or because they instantiate some inherently meaningful type, or because they have been willed to mean something (and not some other thing) by some self-present subject (remember that Hirsch equivocates between the type and the authorial will explanation); instead, they mean contingently and fleetingly, on the basis of their ever-shifting relationships with other words in the language, that is to say, on the basis both of what they are not, and of what they could, in other circumstances, be or mean.

Where Derrida goes beyond the Saussurean analysis of language as a system of differences is in his introduction of the temporal element of deferral. As Christopher Norris explains, 'this involves the idea that meaning is always *deferred*, perhaps to the point of an endless supplementarity, by the play of signification', and this continual substitution of one term for another cannot be either determined or predicted in advance by any writing subject (2002: 32). It is 'not simply a metonymical operation' but rather alters the 'conceptual identities' of the signs that it circulates and substitutes; neither can it be arrested, except strategically and temporarily, by any reading subject (Derrida 1981b: 14). Thus meaning is never fully achieved, fully present, in the way that the intentionalist critic would like it to be, because it is always already interrupted by *différance*: 'speech presupposes (and wills itself to forget) the *différance* of writing', but we cannot forget for long (Norris 2002: 47). This is not to say, however, that this process of substitution and disruption and deferral is not, in its own way, purposive. How, then, can we think intentionality in terms of this Derridean take on temporality – specifically, in terms of the deferral, through time, of the meaning that is produced by a written text? How might this deferral, which is simultaneously a 'precipitation' towards meaning and a resistance to meaning-closure, be thought of as intentional?

Derrida interprets the Latin verb *differe*, from which the French *différer* and ultimately Derrida's own *différance* derive (and which has given us in English the two distinct verbs: to differ and to defer), as including within its remit 'the action of putting off until later, of taking into account, of taking account of time and of the forces of an operation that implies an economical calculation, a detour, a delay, a relay, a reserve, a representation' (1982: 8). All of these Derrida summarizes in his employment of the term 'temporization', adding that in its 'deferral' sense, *différer* means:

> To temporize, to take recourse consciously or unconsciously, in the temporal and temporizing mediation of a detour that suspends the accomplishment or fulfilment of 'desire' or 'will', and equally effects this suspension in a mode that annuls or tempers its own effect. (ibid.)

For 'desire' and 'will' we might here substitute 'intention'; specifically, in the case of literary works, the author's meaning-intention which can no longer be seen as 'accomplished', as 'fulfilling' itself in the achievement of an unmediated, a completed textual meaning. The wandering 'detour' that Derrida describes here under the aegis of 'temporization' is not a temporary detour but instead an unending procrastination; it is also, notably, an 'activity', an 'operation' of the text itself. By contrast, the blueprint intention is generally figured as finite in its ideality, its presence: to the extent that it does persist through time, it does so only until it is (more or less) accomplished or realized by the text, and therefore is strictly teleological.

What *différance* does most radically is to dispense with the notion of an intentional agency which is predicated of some subject – although not, as I hope is becoming clear, with the idea of intentional agency altogether (to the extent that it 'is' anything, *différance* is at least active). Hence Derrida's claim that: 'There is no subject who is agent, author, and master of *différance*, who eventually and empirically would be overtaken by *différance*. Subjectivity – like objectivity – is an effect of *différance*, an effect inscribed in a system of *différance*' (1981b: 28). Moreover, the subject 'is not present, nor above all present to itself before *différance*', in fact the subject, in Derrida's estimation, 'is constituted only in being divided from itself, in becoming space, in temporizing, in deferral' (29). What *différance* destroys, then, is the idea of being/subjectivity as pure presence, whether temporal or spatial, and as the origin of all meaning and meaning-intentions; *différance* is his alternative to the 'metaphysics of presence' and the doctrine of the originating idea or form. So *différance*, although preceding the subject, has no origin: 'there is nowhere to *begin* to trace the sheaf or the graphics of *différance*. For what is put into question is precisely the quest for a rightful beginning, an absolute point of departure, a principal responsibility' (1982: 6). For this very notion of a 'point of departure' presupposes something outside writing, some pre- or extra-grammatological concept which will describe or determine writing, and, of course, *il n'y a pas de hors texte*. In the '*Différance*' paper, Derrida merely claims, somewhat less contentiously, that 'no transcendent truth present outside the field of writing can govern theologically the totality of the field', and much of his work, as we have seen, has comprised an attempt to undermine such notions of theological governance, by the author-god or others, of this unmeasurable, all-encompassing 'field' of writing (7).

In so doing he has made redundant the blueprint conception of intention which functions in this way as a kind of separable origin and cause, but he has also refuted any suggestion of the secondariness of writing as product and effect: both intention and text, then, have emerged from his account of them transformed and radicalized. If we nevertheless read him as retaining some notion of intentional agency in his thinking of text, then this must be something which is always already within the general text and which is, indeed, produced by it, to the extent that it has an origin at all. Certainly, Derrida has an interest in retaining some notion of *force*, for in his attack upon structuralism in 'Force & Signification', Derrida cites its dependence upon phenomenology – they are more commonly figured as antithetical – and bemoans the fact that 'One would seek in vain a concept in phenomenology which would permit the conceptualization of intensity or force. The conceptualization not only of direction but of power, not only the *in* but the *tension* of intentionality' (1978: 27). In this way he gestures towards an intentionality of form which is neither unidirectional nor teleological but instead constituted by tension or difference: *différance*. Brooks, as we saw, was very effective in describing the tension of a *plot*, the feints and false starts and deliberate wrong turns which both frustrate and encourage the reader's desire and which also constitute the desire of the narrative itself, with its necessary aporias and (yet) its undeniable momentum towards some unspecified and perhaps unreachable end. Brooks' account emerges from a specifically Derridean understanding of textual force; might it not then be possible to think of this 'desire' and this 'tension', of these movements to and fro, in terms of the meaning production of the text, rather than that text's plot? Intentionality, on this view, would be a particular recognition of, as Derrida indicates earlier, both the *directedness* (in) and the *power* (tension) of a text: the fact that it creates meaning, endlessly – the mark is a 'machine', remember – and yet simultaneously deconstructs, undermines, undercuts its own supposedly central themes and messages and meanings, in a movement to and fro which we, as readers or critics, may be able to track, but can never absolutely pin down. This is a movement (of meaning production and elision) which is never brought to anything more than a strategic and temporary halt (hence the emphasis on temporal deferral); it is also a movement which can be identified at the level of the sentence (think of one of Henry James') or even at the level of the word (where all the different possible resonances of that word, past, present and future, are put into play). It is in his identification of the 'tension' of intentionality that Derrida moves away from Husserl (this is a directedness which turns back upon itself, a momentum that is never straightforward) and away also from the New Critical understanding of the text as organic unity; the emphasis on tension is also reminiscent of Blanchot's (1982) account of the reading process as *agonistic* – this is a tension which will never be resolved.

So, then, if the key motifs of *différance* are movement and temporization, then my submission is that we might see in this movement of deferral in particular something purposive: an intentionality which is not, finally, of the subject, but which constitutes that subject, and which is evident in the play and openendedness of textual meaning (where 'textual', as Derrida's concept of archiwriting shows, indicates more than merely a written text). Derrida himself is characteristically equivocal when it comes to describing the extent to which *différance* is active, remarking that the 'playing movement' of *différance* which produces differences (briefly, meaning substitutions) is 'not simply an activity' (1982: 11); in a similar vein, he observes that the -ance ending in French, which is derived from the present participle ending -ant, 'remains undecided *between* the active and the passive' (9). He seems constantly to be denying the intentionality of any 'activity' of *différance*, while simultaneously re-deploying the motifs of movement and direction and production and generation which themselves suggest intentionality – in the sense of being-intent-upon something and in their creation of the significant tension described earlier. What we have here, then, is a re-thinking of phenomenological intentionality stripped of its subjective foundation and of its fixed object (its single direction); thus language, although intent upon meaning, is not intent upon any particular meaning, as dictated by any origin or end; its purpose is to be meaningful (to create, in fact, a superfluity of meanings), and it is purposive, and this purpose is something integral to language – or, more accurately, in Derridean parlance, to the gram or mark.

Is it, however, possible to be purposive, as I am suggesting the movement of *différance* is, without being either explicitly teleological or totalitarian in a way that Derrida would evidently find problematic? The emphasis on tension can certainly be read as an attempt to alleviate such concerns, and it might also help here to look at what Derrida implies by the 'dissemination' of meaning which is tied in to this movement of *différance*. 'Dissemination' is not equivalent to 'polysemia', for the latter occurs, like 'metaphor' traditionally taken, 'within the implicit horizon of a unitary resumption of meaning', that is, is part of 'a teleological and totalizing dialectics that at a given moment, however far off, must permit the reassemblage of the totality of a text into the truth of its meaning' (1981b: 45). In other words, it serves to preclude any further wandering. By contrast dissemination 'marks an irreducible and *generative* multiplicity' – again, there is this suggestion of power and productiveness – 'the supplement and the turbulence of a certain lack' – again there is the emphasis on an absence, a non-presence which structures the mark: 'The supplement and the turbulence of a certain lack fracture the limit of the text, forbidding an exhaustive and closed formalization of it, or at least a saturating taxonomy of its themes, its signified, its meaning' (ibid.). The movement and open-endedness of the intentionality which structures the text dictate that the process of interpretation is never finished, is for the moment only, part of a critical context which will likewise never be 'exhaustively' determined or determinable.

We could rightfully object that there is little critical or methodological advantage to this model of textual intentionality, although it does, at the very least, say something to us about *what texts do* and *how they mean*, and what they therefore require of us as readers: namely, a recognition of this tension and desire which will largely produce, structure and characterize our own reading experience, which will, in fact, constitute us as reading subjects. Above all, what should have become clear here is that, despite his apparent anti-intentionalism, it is not the case *either* that Derrida utterly absolves us of the need to think about intention *or* that it is a concept not to be found within his worldview.

Coda: de Man, intention and the inhuman

The grammatological intentionality which I have attempted to identify in Derrida's account of writing and *différance* presents us with an intention which is predicated of no subject, which is rather the generative movement of signifiers within a general text. This 'generation' of meanings is, at times, expressed in a metaphor of mechanization by Derrida, so we would not go far wrong in noticing an anti-humanism in any intentionalism that we might derive from him; at the very least it is identifiable as anti-subjectivism. Analogously, Paul de Man, in his work on meaning and translation, conveys something of the inhumanity of language and sign systems generally. This 'inhumanity' can be read as the extent to which sign systems exceed and repel the governance of the subjective intention, but as de Man so pertinently identifies, this conception of intention as 'subjective' (of or pertaining to the subject) is a limited and flawed one.

In assessing the New Critical treatment of 'form' and 'intent', he sees Wimsatt and Beardsley and others as choosing the former over the latter, as if the two were mutually exclusive – this polarization of form and intent is, of course, the way of thinking that I have been attempting to debunk by degrees in the course of this book. In choosing to do this, the New Critics decide to view the text as an object rather than an act, and thus to assert its formal independence from any mental act or state of the author's. De Man describes Wimsatt as wanting to:

> defend the province of poetry against the intrusion of crude deterministic systems, historical or psychological, that oversimplify the complex relationship between theme and style [. . .]. [Wimsatt] focuses on the concept of intention as the breach through which these foreign bodies reach into the poetic domain. (1983: 24)

So it is intention which is seen to threaten the autonomy of the work, rendering it harder to analyse as a 'totality' or unified structure. But in thinking this way Wimsatt mistakes both the nature of the text (which does not submit to the either/or of the act/object distinction) and, more crucially, the nature of intention, which is unquestioningly figured as 'mental content'. It is worth quoting de Man's objection in full, as it appears to me to be decisive in establishing his carefully nuanced understanding of both text and intention, and it also ties in well to Derrida's deconstruction of the mind/matter and force/form oppositions, as previously discussed. The New Critical suppression of the intentional character of literature rests, de Man claims:

> on a misunderstanding of the nature of intentionality. 'Intent' is seen, by analogy with a physical model, as a transfer of a psychic or mental content that exists in the mind of the poet to the mind of a reader, somewhat as one

> would pour wine from a jar into a glass. A certain content has to be transferred elsewhere, and the energy necessary to effect the transfer has to come from an outside source called intention. This is to ignore that *the concept of intentionality is neither physical nor psychological in its nature, but structural*, involving the activity of a subject regardless of its empirical concerns, except as far as they relate to the intentionality of the structure. The structural intentionality determines the relationship between the components of the resulting object in all its parts, but the relationship of the particular state of mind of the person engaged in the act of structurization to the structured object is altogether contingent. (25; my emphasis)

It is this 'misunderstanding' which has skewed the course of the intentionalism/anti-intentionalism debate since the first publication of 'The Intentional Fallacy' in 1946. Intention has consistently been viewed as 'an outside source' (think of Wimsatt and Beardsley's 'external' evidence) or a 'psychic or mental content' which must be 'transferred elsewhere' (put into the text by the author, e.g., as a kind of illocutionary addition, or perhaps by the reader); the difficulties attending such a 'transferral' and the alleged inaccessibility of this 'content' have time and again lent fuel to the arguments of the anti-intentionalists. My contention is, and has been, that only an understanding of intention as 'structural', such as de Man presents us with here, will facilitate a fruitful reconsideration of the question of the intentionality of literary meaning, and so move the debate forwards.

De Man goes so far as to submit, somewhat contentiously given its influence on subsequent formalisms, that 'the partial failure of American formalism [. . .] is due to its lack of awareness of the intentional structure of literary form', its tendency instead to reify the text as a kind of natural object, something 'organic', rather than the intentional object that de Man claims it is (27). This conception of intentionality as 'structural' seems to me to derive from a phenomenological understanding of intentionality as a system of relations and a kind of directionality, rather than as some kind of mental 'content' or as the transferral of this content. As has been extensively documented, it is the transfer-of-content metaphor which alienates intention from the text which it purportedly intends by insisting on this 'crossing', this perilous movement from one 'place' to another, or alternatively on a transformation of one thing into another (authorial mental state into public, legible mark or into some effect in the mind of the reader); with *trans*ferral and *trans*formation comes the possibility of *trans*gression (which is, in this instance, a corruption or loss of the original 'content').

Like Derrida, however, de Man is quick to stress that literary language displays a particular kind of directed but non-teleological intentionality, thus 'the aesthetic' is described as a case of 'aim-taking for its own sake' – it is the kind of act (he makes an analogy with a hunter aiming at an artificial target) which 'reflects back upon itself and remains circumscribed within the range of its own intent' (26); although this is perhaps problematic in asserting the separability of the 'aesthetic' and the non-aesthetic in a way that Derrida would doubtless take issue with. In addition, de Man succumbs to a rather questionable logic of surface and depth in his critique of formalism, in declaring that the standard formalist rejection of intentionality leads to 'A hardening of the text into a sheer surface that prevents the stylistic analysis from penetrating beyond the sensory appearances to perceive this "struggle with meaning" of which all criticism, including the criticism of forms, should give an account' (27). Although situating intention within the text, the necessarily hierarchical language of surface and depth – intention being a case of the latter – is problematic, because it still represents intention as something mysterious, buried and therefore potentially inaccessible, and criticism as a definitively archaeological process. The suggestion of a 'struggle with meaning', however, hints at a tension within the text (compare Derrida's insistence on the 'force' which disrupts form, and his idea of in-*tension*-ality) which is figured as enacting *its own* struggle for meaning, a struggle which arises, perhaps, out of the conflicting intentions of the marks of which a work is comprised – that is, the different 'directions' in which each mark points or leads us, the different possible meanings that each has or could intend. This struggle means that for de Man too the work is never truly complete: taking into account the Heideggerian emphasis on temporality and becoming, de Man asserts of literary works that:

> The form is never anything but a process on the way to completion. The complete form never exists as a concrete aspect of the work that could coincide with a sensorial or semantic dimension of the language. It is constituted in the mind of the interpreter as the work discloses itself in response to his questioning. But this dialogue between work and interpreter is endless. (31–32)

This owes a great deal both to the phenomenology of reading propounded by Poulet, in its highlighting of 'process' and dialogic becoming, and to an understanding of Derridean *différance* as unending generativity (and the tension of a text which is never resolved), and it also posits a kind of formal incompletion at odds with the New Critical conception of the text as totality.

Elsewhere, de Man's thinking on language and intention acquires a slightly different slant. In *The Resistance to Theory* (1986), de Man approaches the subject of the inhumanity and/or autonomy of language via an analysis of Benjamin's essay on translation, 'The Task of the Translator' (published in *Illuminations*, 1999). In particular he interprets Benjamin as suggesting that the stability and 'canonicity' of the original text is challenged by the translation, and that in literary criticism a similar process unfolds, where, 'the original work is not imitated or reproduced but is to some extent put in motion, de-canonized, questioned in a way which undoes its claim to canonical authority' (1986: 83).

The original, indeed, is revealed by the translation as 'always already disarticulated' or fragmentary and this implies, on de Man's reading, a general will towards disarticulation existing within language, a kind of negative intentionality in fact (84). The work is intent upon its own semantic dissemination. What is most disturbing about translation is the way in which it presents us with 'the fiction or hypothesis of a pure language devoid of the burden of meaning' – presumably because it purports merely to replace the language of the original with the language of the translation, without reference to any prior or posterior or other meaning (the Idea, the self-present intention), and without itself creating any new meaning (84). The will-towards-disarticulation of language imposes on us a particular kind of 'suffering' or alienation, according to de Man's reading of Benjamin. More specifically, the disarticulation occurs in the disjunction between '"das Gemeinte", what is meant, and the "Art des Meinens", the way in which language means', which de Man translates as the difference between 'vouloir dire' (to mean) and 'dire' (to say) in French, and which we might be tempted to translate into the disparity, potential or actual, between intention or speaker's intended meaning and linguistic or lexical meaning (86).

In fact it is more complex than this. The English translator of Benjamin's 'The Task of the Translator', Harry Zorn (although de Man rather perplexingly refers to him throughout as 'Zohn'), translates 'das Gemeinte' and 'Art des Meinens' as 'the intended object' and 'the mode of intention', respectively; Benjamin's claim is thus that although 'the words *Brot* and *pain* "intend" the same object [. . .] the modes of this intention are not the same' and the words have different resonances and produce different effects in their respective languages, despite apparently 'intending' the same object, bread (Benjamin 1999: 75). It is this disparity which makes genuine *translation* (as opposed to *transformation*) impossible.

From Benjamin's assertion that 'all translation is only a somewhat provisional way of coming to terms with the foreignness of language', comes de Man's more radical proposition concerning the inhumanity of language (ibid.). This is because de Man takes issue with the 'phenomenological assumption' underlying Zorn's translation, the assumption, following Husserl: 'that [both] the meaning and the way in which meaning is produced are intentional acts' (1986: 86–87). In contradiction of this, de Man asserts that:

> It is not *a priori* certain at all that the mode of meaning, the way in which I mean, is intentional in any way. The way in which I can try to mean is dependent upon linguistic properties that are not only [not] made by me, because I depend on the language as it exists for the devices which I will be using, it is as such not made by us as historical beings, it is perhaps not even made by humans at all. (87)

De Man appears here to be developing the intention versus convention problem which so troubled Hirsch, to the point of expelling human agency from the realm of language altogether. At best, he signals the impotence of the subject – 'I can try to mean' – in attempting to delimit the meaning of his words. Like Derrida he is suggesting that we are within language to the extent that our meaning-intentions and their effects are localized; they do not precede or produce language as a system, neither can they preclude *what* we mean being 'upset' by the *way* in which we mean, which is to say, by the materiality of language, its consequent independence of us, and the unwanted significances and connotations that a word/phoneme can and will introduce:

> If language is not necessarily human – if we obey the law, if we function within language, and purely in terms of language – there can be no intent; there may be an intent of meaning, but there is no intent in the purely formal way in which we will use language independently of the sense or the meaning. (ibid.)

The distinction which I think de Man is making here is between an intentionality of meaning, that is, of the words themselves (which he concedes) and an intentionality of use, that is, a governing intention of the author or speaker (which he dismisses, and which is vital to the traditional intentionalist argument); he also subsequently explains it as a distinction between hermeneutics and poetics and this can be tied in to a common theme of de Man's, namely, the rhetoricity of (literary) language as a kind of agency built in to language.

It is this agency/rhetoricity which permits – even induces – language to evade the control of the subjective intention, to miss the mark and go off in unexpected, unintended directions. In the *Blindness & Insight* (1983) collection of essays, he explores the idea of a 'blindness' to the true 'insights' (or meanings) of a text, as something fundamental to writing and this is, notably, an authorial blindness, such that the author is shown to be often ignorant of the meaning which his language intends. Thus he argues that 'the rhetorical character of language opens up the possibility of the archetypal error: the recurrent confusion of sign and substance', adding further that 'it follows from the rhetorical nature of literary language that the cognitive function resides in the language and not the subject', bringing us back again to this suggestion of the 'inhuman' – or, at least, non-subjective – character of meaning, while also carrying echoes of Poulet's attempt to treat the literary work as a kind of subject in its own right, displaying a sort of intentionality (136, 137).

What matters to de Man, then, is not the 'blindness' or 'insight' of the author in relation to 'his' text, but rather the question of 'whether his language is or is not blind to its own statement' (ibid.); he is interested in how the language reflects upon and is aware of itself, its own portents and pitfalls. Elsewhere – in *Allegories of Reading* (1979) – de Man moves beyond his analysis of literary language to claim that all language is figural: 'the figurative structure is not one linguistic mode among others but it characterizes language as such' (105) and this figural character of language means that misreading is always possible, is indeed 'a basic condition of producing meaning at all' (McQuillan 2001: 35). So, although an author or speaker may 'intend' a meaning in one sense, by meaning to produce it or willing its production, in another sense he cannot 'intend' it: he cannot dictate its directionality or scope (which, in all likelihood, will not cohere with this original meaning-intention and which will also undermine and turn back upon itself at each moment).

When challenged to justify the strength of his 'inhumanity of language' claim, in questions following the delivery of his paper on 'The Task of the Translator' (the transcript of which question and answer session is included in *The Resistance to Theory*), de Man implies that what he means by 'inhuman' is in fact 'unintentional' (in this sense of an intention which is strictly of the subject):

> The 'inhuman' [. . .] is not some kind of mystery, or some kind of secret; the inhuman is: linguistic structures, the play of linguistic tensions, linguistic events that occur, possibilities which are inherent in language – independently of any intent or any drive or any wish or any desire that we might have. (1986: 96)

But despite stripping language of intention in this (blueprint) sense, he does allow for what he describes as semantic intention or intentionality (de Man tends to use 'intention' in the phenomenological sense of being-directed-upon), which connotes the way in which language is 'oriented to meaning' or 'intends meaning' (94). This 'intending', then, is something which has been taken out of the hands of individual human subjects and which is now predicated of language itself; to the extent that it is meaningful, de Man affirms, it will always be intentional. In this way he appears to answer in the positive his questioner Billy Flesch's suggestion that you could 'get rid of the subject without being left only with language', that is, a language stripped of intention (ibid.).

De Man's remarks about the relationship between intention and language and the subject are characteristically abstruse, and nowhere does he develop this hint regarding the intentionality of language into a more comprehensive linguistic philosophy, perhaps because the rhetoricity of language would lead him to doubt the very comprehensiveness of this in the first place. Nevertheless, even these few remarks point to a willingness to explore other conceptions of intention/intentionality than that offered us either by traditional intentionalist accounts of meaning, or by phenomenology. Furthermore, these other conceptions of intention, as both Derrida and de Man insinuate, must resist the urge to tie it either to some human subject (as source) or to some singular meaning (as end point or object).

The advantages of Derrida's 'grammatology' seem to me to be precisely the ways in which his attack on origins, on temporal and spatial presence and on the perceived supplementarity or secondariness of the text finally abolishes the blueprint conception of intention and the traditional authorial intentionalism that has relied upon it. There is another practical advantage to his work in the way that *différance*, because of its spatio-temporal character – specifically, the way in which it deconstructs the space/time distinction by suggesting both 'spacing' (differing) and 'temporization' (deferral) – is both formalist and historicist, and thus allows Derrida to escape the rigidity of the critical either/or. In somehow extricating himself from this critical logic, Derrida opens up the possibility of a new and expanded variety of formalism which need not reject intention as paradoxically 'exterior' (outside or prior to the work) in its immaterial interiority (its presence-to-mind as pure 'expression'); for now, post-deconstruction, the outside is the inside, and vice versa, and form and force are inextricable in a way that had not previously been imagined. What I hope this means is that we may not need to dispense with intention altogether – this is my central claim – but we do need to dispense with any notion of the text as a straightforward expression of or communication from the subject.

It is these expressivist theories, not intention, which are Derrida's targets, for he observes that 'communication' as a concept

> implies a *transmission charged with making pass, from one subject to another, the identity* of a *signified* object, of a *meaning* or of a concept rightfully separable from the process of passage and from the signifying operation. Communication presupposes subjects (whose identity and presence are constituted before the signifying operation) and objects (signified concepts, a thought meaning that the passage of communication will have neither to constitute, nor, by all rights, to transform). (1981b: 23)

In this idea of communication, then, is implied the self-presence of subjectivity, as the origin of meaning, and a 'thought meaning' – an intention, in fact – which is complete, ideal, which precedes any process of signification and which can then undergo that process of substantiation, that making-flesh or incorporation, which is inscription. All of this metaphysical baggage Derrida wants to do away with, while also recognizing that he cannot write outside of this tradition, that he must partake of it to some extent. What he does, arguably, retain is some notion of intentionality, considered partly as purposiveness, a generative movement marked by its internal tension and its resistance to closure, and partly as directionality, but a 'wandering', disruptive, directionality that evades some final *telos*.

Does this, finally, come down to a choice to be made between language and the subject as the origin of meaning and owner of intention? In his reading of Georges Poulet, de Man claims that, for Poulet:

> Self and language are the two focal points around which the trajectory of the work originates, but neither can by itself find access to the status of source.

> Each is the anteriority of the other. If one confers upon language the power to originate, one runs the risk of hiding the self. This Poulet fears most of all, as when he asserts: 'I want at all costs to save the subjectivity of literature.' But if the subject is, in its turn, given the status of an origin, one makes it coincide with Being in a self-consuming identity in which language is destroyed. (1983: 100–101)

In speaking of literary intention, I would suggest that this choice between the subject and language *does* need to be made, and that this is analogous to a choice between intention, traditionally conceived as mental, present, finite and subjective (of the subject), and intentionality, as something potentially predicated of structure or form or as a relation of formal elements to each other, implying both direction and agency or generation. In broader terms, literary theory of the late twentieth century has chosen language over the subject, has moved, then, from a philosophy of subjectivity to one of textuality. Deconstruction, as the most prominent proponent of the (anti-)philosophy of textuality does not, however, dispense with the subject altogether; but it does attribute to text, in its most general and expansive form, an intentionality and agency of its own which relegates the subject to the position of textual product or effect.

What is most taxing is to conceive of this structural intentionality as somehow non-totalizing, not a relationship of parts to some completed and closed whole. Ultimately, I am not arguing that either Derrida or de Man present us with a ready worked-out theory of the intentionality of form, only that the radicalization of form and text that they perform requires us to rethink what we might mean by intention and intentionality accordingly, and how we might distance these concepts from an outdated model of subjectivity. This radicalization of text involves the attribution of an agency – and in de Man's case even a suggestion of cognition – to the text, which may or may not owe something to their earlier dabblings in phenomenology, although mine is not, strictly speaking, a genealogical account. I am arguing, therefore, that their work does not militate against such a rethinking of intention, and in this sense they are not the anti-intentionalists that they might at first glance appear to be (although they could hardly be exempted from the accusation of anti-authorialism). In my eyes, the radical potential of their work which has been as yet unrecognized, is the suggested separation of intention and the subject (thanks in great part to the separation of the text and the subject). Crucially, then, this separation of intention and subject – *intention after the subject* – need not amount, as the legacy of deconstruction (and, indeed, of decades of anti-subjectivist and anti-intentionalist theories of literature), to a violent and irrevocable sundering of intention from text.

Note

1 I say 'of necessity' because this sundering is seen as ensuring its transcendence, thereby ensuring, importantly, that it will not be corrupted by the potential semantic flux of the text. Paradoxical this: that the immaterial – the oft appealed to mental blueprint – should be seen as substantial enough to guarantee the meaning of the material, and yet semantically insubstantial and unstable text.

Conclusion

The Ethics and Pragmatics of Intentionality

It has been the work of this book to counter what I have identified as a kind of double exclusion of intention: first, its exclusion from the text, a text which repels it and situates it as prior, as the outside to its inside, the force which must be added to its form; second, the exclusion of the very term 'intention' from our critical lexicon in recent years, such that it has become practically a taboo word, one which invokes all kinds of troubling associations and even implies a stance that is resolutely anti-theory or which harks back to some supposedly pre-theoretical critical mode (the very existence of which seems to me to be extremely doubtful). The recuperation of intention that I have attempted has therefore had a correspondingly double purpose: it has involved bringing intention back 'within' the text, while simultaneously and necessarily worrying away at this inside/outside distinction in an attempt to prove its essential arbitrariness and permeability; and it has also endeavoured to bring the term 'intention' back within our critical vocabulary in a way that does not treat that vocabulary as inherently suspect or hostile (to literature and to the discipline of literary studies) – for it seems to me that it is both possible and desirable to advance a version of intentionalism that is not radically at odds with, for example, the language and thinking of post-structuralism.

What has resulted is the sketching out, in a preliminary fashion, of a theory of the intentionality of literary form – a theory facilitated by, although not expressly embodied in, the linguistic philosophy of Jacques Derrida and suggested also by certain isolated but pertinent comments of Paul de Man on intention and subjectivity. Deconstruction has created an *aperture* for intention through its radical re-thinking of literary form, and writing in general – the latter is reconceived as a generalized, expanded and expansive writing, as Derrida explains:

> [While one used to say] 'language' for action, movement, thought, reflection, consciousness, unconsciousness, experience, affectivity, etc. Now we tend to say 'writing' for all that and more: to designate not only the physical gestures of literal pictographic or ideographic inscription, but also the totality of what makes it possible [. . .], we say 'writing' for all that gives rise to an inscription in general. (1976: 9)

In fact, 'writing' can now be said to 'exceed' and 'comprehend' 'language' (8), the latter becoming a sub-category of the former. Derrida here exploits the malleable meaning of 'supplement' in French, to suggest both that which is additional and that which usurps or produces whatever it is supposed to be added to. His new concept of '(archi-)writing' (1981b: 8) or 'primary writing' (1976: 7), consequently and simultaneously, 'provokes the overturning of the hierarchy speech/writing, and the entire system attached to it' (1981b: 42), and this 'overturning', note, is not a simple reversal, but one which 'releases the dissonance of a writing within speech, thereby disorganizing the entire inherited order and invading the entire field' (ibid.). The aural metaphor – 'dissonance' – that Derrida employs here also works at blurring the speech/writing distinction. The idea of dissonance 'invading' a 'field' also neatly avoids a simplistically spatial analysis (the 'field' in question becomes an indeterminately bordered force field, or even the period for which an echo hangs in the air, rather than a measurable space). In dismantling these metaphysical oppositions (space/time, speech/writing), Derrida is removing those obstacles which have hitherto precluded a conception of writing as intentional by positing it as something separated from – and therefore requiring some injection or addition of – force; Derridean archi-writing is no longer separated from force, or from intentionality figured as a kind of animating force.

Where we take this idea of textual intention (an intention *in* and *of* writing) from here, what the effects – if any – upon critical practice will be, must be the subject of a further work, but it might be useful nonetheless to begin to speculate here on what exactly is meant by the 'intentionality of form', and what the ramifications of such a theory could or should be.

Textual intentionality is a formal or structural feature of texts. It is therefore possible to analyse texts in terms of the intentionality that they themselves display without having recourse to the imaginative input of authors or the emotional responses of readers, for example. 'Intentionality', building upon the use of this term by phenomenologists such as Husserl and Brentano, here means a kind of directedness: literally the way in which certain words, phrases and sentences, and groups of sentences in their combination or juxtaposition, 'intend' meaning and so, in consequence, send the reader in all sorts of different directions. Peter Brooks' understanding of the dynamism of plot is useful here, particularly his suggestion that the reader must negotiate the 'field of force' which falls between the beginning and (nominal) end of a narrative (1984: 47); where he thinks of this 'field of force' predominantly in terms of plot, I am attempting to think of it in terms of a level of linguistic meaning which is (tentatively) more basic than plot – the level of meaning out of which plot is produced.

If formalism is taken as a critical model, then it might at first appear that the movement of intentionality within a text is something that emerges in a quite straightforward fashion from the relationship between the parts of that text, so we can raise for consideration here the claim that *textual intention is to a certain extent concerned with the relationship of parts to whole in a text.* This is a way of seeing the text that emerges from earlier formalisms such as Russian Formalism and Structuralism, which conceive of the text as embodying a code and possessing a determinable structure; as a result the text is seen as analysable in terms of the 'functions' of its different parts or units insofar as they contribute to this structure. This type of analysis gains an extra dimension through post-structuralism, which is really an injection of scepticism regarding the determinability of the codes and conventions according to which texts operate. Thus the relationship of parts to whole is attended to – and it is possible to read this relationship as demonstrating a kind of internal purposiveness of the text – but this must be accomplished with a concomitant recognition of, first, the shifting and contingent nature of this relationship (or relationships, in the plural) and, second, the essential arbitrariness and openness of this 'whole' or the irreducibility of its 'form'.

The intentionality of a literary text can therefore more accurately be described as supervenient, in a particular way, upon the intentionality of language itself; it is an agency at the level of the mark. Although this understanding of a text invites a critical approach which may justifiably be termed 'formalist', as its main focus is the 'text itself', the very term 'formalism' must be understood in relation to the deconstructive account of form which subjects this concept to a thoroughgoing critique. For Derrida, 'form' is, inevitably, metaphysical in its meaning and associations:

> As soon as we utilize the concept of form – even if to criticize *an other* concept of form – we inevitably have recourse to the self-evidence of a kernel of meaning. And the medium of this self-evidence can be nothing other than the language of metaphysics. In this language we know what 'form' means, how the possibility of its variations is regulated, what its limit is, and in what field all imaginable objections to it are to be contained. (1982: 157)

The concept of form, in Derrida's estimation, 'cannot be, and never could be, dissociated from the concept of appearing, of meaning, of self-evidence, of essence. Only a form is *self-evident*, only a form has or is an *essence*, only a form *presents itself* as such' (158). So in any use of 'form' there is always a reference to 'the theme of *presence in general*', leading Derrida to announce that 'form is presence itself' (ibid.). However, the introduction of force into the equation, such as Derrida attempts in 'Force & Signification' (1978), presents us with an idea of form as deferred and disrupted presence – rather than as some kind of static and finite and complete structure – and this is how I am attempting to think of a literary text: as form through which a certain force works itself out, endlessly, a force which it is possible to describe as 'intentional'.

A further proviso here, then, is that *textual intention is premised upon the inextricability of form and force.* What has become evident in the course of this book is that the relationship of form and force has proved problematic for intentionalists and anti-intentionalists alike, and critical approaches to literature tend to assign themselves the task of investigating either one or the other. This is despite the fact that the separation of form and force is always fraught with difficulties: witness the various attempts, under the aegis of speech act theory, to institute a firm distinction between locutionary and illocutionary meaning. Knapp and Michaels' answer to this problem (although not presented in the language of speech act theory) is to assert that there is, effectively, no such thing as 'bare' locutionary meaning: either words are animated by some addition of illocutionary force which renders them meaningful or they are not meaningful and not, strictly speaking, words at all. Hirsch, too, struggles with the reconciliation of intention (conceived of as the personal injection of force that the author puts into a text) and convention (figured as literal or linguistic meaning, the formal type-meanings that the author must choose from).

These thoughts about form also, helpfully, allow us to assert that *textual intention is material – or at least premised upon an understanding of the materiality of the text.* This runs contrary to the view of the text offered by Poulet, who sees the text as transcending its material status in order to become meaningful for the reader, so books 'wait for someone to come and deliver them from their materiality' (1969: 53). Yet, for Derrida it is the very materiality of the mark – its dogged material existence through time – what Searle more prosaically terms its 'permanence' (1977: 200), which ensures it both a life and a form of agency beyond that of the author whom it must leave behind; this agency is at least in part the result of it endlessly being situated in new contexts, contexts which are not and cannot ever be exhaustively determinable; in this way, what Searle reads as an unproblematic 'permanence' becomes, for Derrida, a necessary 'impermanence' or instability of meaning, an open-endedness of the text.

Traditionally, the materiality of writing has been taken as evidence of its secondary or supplementary character and 'intention' has been interpreted as something ideal or immaterial (mental content) which must somehow be instantiated by the text; the text represents the intention – in this way the two are kept apart. Derrida's breaking down of the mind/matter distinction – his interrogation of its necessarily hierarchical nature within a phonocentrist way of thinking – and his re-reading of supplementarity, again allow us to move beyond a notion of intention as mental content awaiting incorporation in some set of physical marks. Instead, intentionality is already part of the materiality of the text and this materiality does not mark writing out as 'exterior' (by contrast with the supposed 'interiority' and self-presence of thought); in fact, Derrida's description of archi-writing makes it not only material or physical, as the previously cited passage shows, but rather 'writing' comes to stand for 'the totality of what makes [inscription] possible', thereby blurring the boundary between material and immaterial (1976: 9).

The deconstruction of the mind/matter distinction which Derrida performs is also symptomatic of a more general interrogation of the relationship of inside to outside, and this gives rise to the memorable declaration in *Of Grammatology* that 'the outside is the inside' (1976: 44). The imminent collapse of this distinction has surfaced at various points in the theories discussed in the preceding chapters: when Wimsatt and Beardsley stutter over and so compromise their own categories of 'internal' and 'external' evidence; when Hirsch's 'meaning' and 'significance' threaten to bleed into one another in a manner which endangers the stability of both classifications. What becomes evident in both cases is the impossibility of keeping intention either totally inside or totally outside a text; intention, more than anything else, violates and so undermines the boundaries between text and world, and between mind and matter. It may be precisely this unsettling or destabilizing quality of intention which makes it simultaneously vital to our understanding of literature (we cannot simply exclude it from the equation, it will always reassert itself at some crucial juncture) and also a kind of threat to that understanding (we cannot pinpoint exactly where in the equation it falls or what its role is, but that role will always involve the troubling of categories and distinctions such as inside/outside or form/force). Why else has there been such a notable accumulation of anxiety around this concept of intention and its placing in relation to the text? The answer that I hope has emerged here is that any consideration of intention forces us to a realization of the complex interdependency and reciprocity of word and world and so to an interrogation of the nature and delimitation of the text, the function of representation, the construction of meaning and other matters which may seem to go beyond the anticipated scope of a purely 'literary' criticism. More ominously, perhaps, this analysis suggests that these questions and these shifting relations will not and cannot be resolved or even comprehensively described – although the result of this is not, I think, to suggest that 'theory' is either pointless or impossible.

One of the advantages of developing a theory of textual or formal intentionality seems to me to be that it goes some way towards describing the peculiar temporality of a literary text, so the next claim to be defended is that *textual intention is concerned with the existence of the text through time.* In the course of my discussion a certain tension has been evident between descriptions of the text as act and descriptions of it as object; what both of these descriptions tend to elide, however, is the nature of the text as temporal artefact, by focusing either on the (completed) act of creation or on the (finished) poem on the page, for example. This has had some bearing too upon the discussion of intention, as critics have tried to locate the meaning-intention in time, usually deciding to situate it prior to the text which it intends. The blueprint account of intention also ends up having to posit the finiteness of that intention, without being able to determine at what point it is 'completed' – before the text is written? Before that act of writing is completed – or only when it is completed? The wilful separation of intention and text thus presents the theorist with the problem of two time schemes to be reconciled: the time of the intention and the time of the text, with the suggestion that one must be mapped onto the other, or that there must be some essential and productive overlap of the two. Accounts of literature which have addressed themselves to the reading process – an ongoing process, a repeated process – have been more successful in capturing a sense of the text as, necessarily, incomplete, but most reader-response theories have thereby sidelined all questions of intention as apparently irrelevant. Among intentionalists, Richard Shusterman alone appears to advance the idea that 'the necessary meaning-securing intentions could belong to readers of the text rather than its original author', but this still has the negative effect (in my estimation) of tying intention to some specifiable subjective consciousness (1992: 67).

What the theory of the intentionality of literary form recognizes, then, is the continuing production of meaning that a text lends itself to; so, as an adjunct to the statement about temporality, it is also correct to say that *textual intention is to do with the production and proliferation of meaning(s) through time.* A literary text is therefore neither act nor object exclusively, but rather a kind of 'activity' in itself: an activity of continual meaning-production over time, with no end point, no closure or finalization of meaning in sight; intentionality of form, then, actually works against the idea of either the text or the intention's completion and so has no need to pinpoint when this completion might occur. It is on the basis of its active and continual meaning-production that we can describe the text as 'intentional': it has the kind of dynamism which Brooks finds in plots (a dynamism derived from this more fundamental dynamism and productivity of language), but also the open-endedness that Ray, in his reading of Blanchot, sees as the necessary concomitant of the description of a literary text 'as a writing intention' (1984: 14); as such it is never completed, or is only nominally and strategically completed by each new reading of it (which only means that the reader's experience of it is completed, for the time being, not that the text itself is 'finished'). This is where textual intentionality departs from the more traditional phenomenological understanding of intentionality, in that it is not intent upon some fixed object (or, in this case, some fixed meaning).

What is useful about Derrida's thinking of *différance*, and why it has played so pivotal a role in the sketching out of this theory of intentionality, is the way that it brings together the idea of meaning-production or proliferation and the idea of the movement of temporality within/by writing. In doing this it moves beyond the act/object distinction, as it forces an encounter between the spatiality and the temporality of the text – a text which thus becomes the site of a differing and deferring which are not merely metonymic and which are, potentially, endless. In particular, *différance* captures the idea of movement which Brooks too puts such emphasis on, so another piece of the puzzle here is the claim that *textual intention is dynamic, to do with motion and movement*; it indicates a quality of agency that the text possesses which arises out of the potential of words, and conjunctions of words, to intend new meaning possibilities and so to shift in relation to each other. *Différance* might also, therefore, seem to have a more or less metaphorical value in what it tells us about the experience of the reader, specifically about the movement of mind that occurs in reading, the process of picking up and following the associations and resonances of each word, of being led off in different directions. However, this shifts the emphasis back to the consciousness of the reader and is in danger, therefore, of suppressing that vital material aspect of the text and figuring it instead as an immaterial experience or affect, a primarily mental phenomenon. For this reason, I would prefer to focus upon the dynamism and intentionality of the text itself, regardless of the difficulty of determining where that text begins and ends, rather than upon the dynamism of the reading consciousness – as those of a phenomenological bent are wont to do.

To underline this point, then, we can say that *textual intention is intrinsic to the workings of written language as Derrida conceives of it* – in other words, marked by iterability and *différance*, both of which concepts are crucial to an understanding of text as dynamic and generative and open-ended. This 'open-endedness', however, requires some further exploration, because the idea of structure that we are being presented with here is exceedingly difficult to grasp, possibly because it asks us to accept textual intention as systematic without therefore regarding it as systematizing – the text is not a unity or totality (as New Critics and Structuralists have implied). Nevertheless, it is possible to read this movement of intentionality as systematic in the way that it constitutes a kind of irrefutable and ineluctable logic of the written; it is systematic in *that* it wills, rather than in *what* it wills (which I suppose amounts to saying that it is systematically anti-systematic – one of those Derridean paradoxes which invites frustration but is not, therefore, meaningless). For the time being it will have to suffice that *textual intention is systematic but not systematizing*.

Another point that has emerged from the preceding chapters is that *textual intention is properly post-humanist*. How could it be otherwise? It is the product of a way of thinking about the subject and about language which sees the former as, to a large degree, produced by the latter – not simply eliminated or elided, but certainly constructed through language, or through 'writing', in the Derridean sense of that term. As such, it makes sense to think this intentionality through metaphors of cybernetic systems (as Christopher Johnson suggests) or machines (as Derrida does) or motors (as Brooks does), because this serves to reiterate the central point that texts intend meanings above and beyond the intervention of any human subject. This is also what I understand Paul de Man to be gesturing towards in his related claims that language itself 'intends meaning' and so exceeds our subjective controls in a way that makes it not human:

> That there is a nonhuman aspect of language is a perennial awareness from which we cannot escape, because language does things which are so radically out of our control that they cannot be assimilated to the human at all, against which one fights constantly. (1986: 94, 101)

Naturally, this argument will have ramifications for our understanding of what is acceptable, or even 'ethical' in the interpretation and wider treatment of a literary text, and I will turn to this matter at length. In the meantime, however, having tried to put together some ideas on what is meant by this term 'intentionality of form' and its correlates, it may also be useful to outline what exactly textual intentionality is *not*, as a means of further distinguishing it from the authorial intentionalism which it sets out to supplant.

Textual intention is not relational (strictly speaking). The intentionality of the text is not something which emerges out of the relationship between text and author, it is not part of the process of reading and understanding, although it certainly has ramifications for that relationship and that process. So this is not to say that the reader does not experience this intentionality, does not come up against the intention of the text in a certain way; it is only to claim that the text is intentional in the absence of any readers, in the same way that it is intentional beyond the intervention of its author or any subject whatsoever; hence the description of this theory as 'post-human'.

Most importantly, *textual intention is not a matter of consciousness.* It is not the product of either the author's or the reader's consciousnesses and it is this 'mentalism', this limiting view of intention as a (more than likely inaccessible) mental state and the problems associated with it, that I have most conscientiously sought to get beyond in the course of this book. The authority which is most spectacularly unseated by the Derridean emphasis on writing as an 'iterative structure' is the authority of the subjective consciousness, that is, of the author of speech or writing. For writing is henceforth, he claims, 'cut off from all absolute responsibility, from *consciousness* as the authority of the last analysis, writing orphaned, and separated at birth from the assistance of its father' (1982: 316). This patricidal impulse of writing is something that Derrida celebrates and, significantly, the loss of this 'assistance' does not thereby render the text meaningless; rather it liberates it. The 'authority' of this originating consciousness is also challenged in itself, in its supposed purity, self-presence and separateness, for writing invades and structures it. Again, this puts to the test our understanding of inside and outside in relation to writing, as writing is no longer 'purely external', for it 'affects memory and hypnotizes it in its very inside' (1981a: 110). The thought or idea does not simply precede writing, there is a two-way relation here which must also have consequences for our conception of intention.

The distancing of intentionality from consciousness and the mental that I have worked towards here also implies that we should not necessarily see this notion of textual intention as involving some attribution of consciousness to the text itself, although it may be convenient to talk about it in terms of will and desire and other mental states or properties. We must consider, though, what the dangers of such anthropomorphism may be. Are we merely replacing one thinking subject (the author) with another (the text)? What are the advantages of retaining this model? Perhaps we can only think of texts as producing meaning or displaying agency or intentionality if we impose upon them this model of consciousness, see them as minds in their own right: Poulet certainly does this, and even de Man, in his assertion that 'the cognitive function resides in the language and not the subject' falls prey to such a tendency (which is, apparently, a tendency to preserve this link between intention and the mental, even where that intention is not predicated of some human subject) (1983: 137).[1] Is this merely metaphorical or does it reveal certain limitations in our thinking of meaning and agency such as, for example, a shying away from the materiality of the text/signifier, as previously discussed? Even if it is 'merely' metaphorical, the Derridean account of metaphor in 'White Mythology' (1982) suggests that such structural and structuring metaphors have the power to alter and/or displace what they are describing. So, despite the appeal of such models and metaphors, I would like to maintain that *textual intention is not dependent upon the presence of some Subject*, and, in addition, that *textual intention is not dependent upon anything 'outside' the text*, while recognizing the precarious nature of this term 'outside', as outlined earlier.

The point of this book, as I have insisted from the outset, has not been to produce a guide to a new way of reading, a comprehensively worked-out critical methodology that future readers may apply to literary texts for guaranteed results (the kind of foundationalist 'theory' that Fish inveighs against so successfully); rather it has represented an attempt to assert the intuitive inextricability of certain concepts (meaning and intention, e.g.) and the possible separability of others (subjectivity and intentionality, e.g.). Nevertheless, I will conclude by gesturing towards what the minimal effects and ramifications of this theory of the intentionality of literary form might be.

In the first place, it asks us to re-think what is meant by 'intention'; initially, to consider what *has* been meant by it, in the various discussions of literary meaning under deliberation here, how 'intention' has functioned in such discussions. As has been shown, this is by no means clear in every case. Additionally, it prompts us to ask what *could* be meant by 'intention' in future, what new connotations it might take on as a result of the unsettling and defamiliarization that I have attempted here. For even a glance at the *Oxford English Dictionary* shows us that 'intention' has carried a number of different senses throughout its etymological history: for example, 'the action of straining or directing the mind or attention to something; mental application or effort; attention, intent observation or regard', *or* 'volition which one is minded to carry out', *or* 'that which is intended or purposed; a purpose, a design' *or*, more intriguingly, 'straining, bending, forcible application or direction (*of* the mind, eye, thoughts, etc.)', *or*, in its 'logical' sense, 'the direction or application of the mind to an object; a conception formed by directing the mind to some object' (*OED* 1989: 1079). 'Intention', then, can be a state, an inclination, an action, a thing; common to most of the definitions is some suggestion of a future-directed aim, so any analysis of intention in a literary text must take into account this temporal aspect and this notion of directedness. Beyond that – and before we even progress to an exploration of the more philosophically complex term 'intentionality' as it is employed (in quite different ways) by phenomenologists such as Husserl, and analytic philosophers of mind and language such as Searle – the way would seem to be clear for the most audacious of redefinitions, wherever this might help to elucidate the workings of literary texts; it may, indeed, be necessary to insist upon a specialist, literary use of this term 'intention', to distinguish it from other, more 'commonsense' (or habitualized) applications. Above all, I have urged a consideration of the possibility of predicating intention of both people and things (subjects and objects) – and in doing this I am harking back to the use of *'intentio'* in Medieval Latin to mean a kind of general effort or agency, as Roskill details (1977).

This theory of the intentionality of form also calls for a rethinking of 'form', of the boundaries of a literary text (or rather, its apparent formal unboundedness) and consequently of our related understanding of force. Much has been said about the formalistic nature of deconstruction, its links to New Criticism, for example, but a great deal more remains to be said about the new and distinctive formalism which emerges from post-structuralism and the way in which it deconstructs the inside/outside opposition. Increasingly, then, we are required to acknowledge the arbitrariness and falsity of the posited choice between formal and sociological approaches to literary texts, by conceding the inextricability of these categories. This may necessitate the overturning of certain critical paradigms, and the introduction of a new theoretical terminology to some extent, certainly one in which 'intention' will not be out of place.

This new terminology must also account for the politics and ethics of such a 'new formalism', a task which I have not even attempted here, despite the obviously polemical stance of this book. One of the accusations levelled against deconstruction has concerned its supposedly apolitical nature or, worse, its inducement to a kind of political apathy among those who consider themselves its adherents, this apathy being figured as the inevitable result of deconstruction's alleged championing of an out-and-out relativism or extreme scepticism. In fact, I don't believe that such a scepticism is argued for, at least not as far as Derrida's earlier works are concerned. And if the kind of rigorous and methodical close readings conducted *by* Derrida and de Man are taken as the model for a more general ethics of interpretation – rather than those more creative and literary readings *of* Derrida and de Man by their self-professed acolytes (Hartman et al.) – then it is certainly possible to see texts as making demands upon their readers, even if this does not involve the commitment to something that we might more conventionally describe as a 'truth'.

What the theory of the intentionality of form implies is that while there is no single 'truth of the text', yet the text may be seen as working out a kind of logic of its own, which the reader is required to apprehend and to follow; the reader may not, then, simply decide the meaning of the text himself, neither is that reading simply a product of the 'interpretative community' of which the reader is, by default, a member. What this logic does not do is determine and so finalize the meaning of the text, but it may yet *rule out* certain meanings and interpretations (this distinction between ruling in and ruling out is a crucial one, I think, and was touched upon in my discussion of Meiland's reading of Hirsch in Chapter 1); the text is not a blank canvas upon which the reader inscribes his own meanings at will. This summons to the reader to comply with the intentionality of the text is a looser conception of ethical criticism than that premised upon some notion of responsibility to the author (such as Hirsch propounds) but it does at least serve to preclude the kind of absolute scepticism about the possibility of understanding that is lamented by Derrida's detractors, and celebrated by his more florid admirers; it is thus more in keeping with the view of deconstruction propounded by Rodolphe Gasché and Christopher Norris – the latter comments, sensibly, that, 'Derrida's scepticism is not what some of his interpreters would make of it, a passport to limitless interpretative games of their own happy devising' (2002: 125). The intentional text, then, is not limitless, properly speaking, despite its potential open-endedness, because it asks that the reader impose upon it a contingent kind of closure, a nominal form; at the same time, it threatens always to exceed these boundaries and to intend new meanings at the next reading, the next glance. Above all, the reader is asked to recognize that this agency with which he must wrestle and come to terms emanates from the text itself and from a latent potentiality of language (as not only a system of differences, but a system which generates and intends meanings – it is in this way that language exceeds the subject without, however, annihilating that subject), not from some real or hypothetical author-figure.

What we owe the text, then, as ethically minded critics is at least our painstaking attention (which is an attentiveness to its intention) and this is an invocation with which Wimsatt and Beardsley would surely not disagree. Ultimately, then, intention and text cannot be separated – and neither one can be considered in isolation – because at every reading of a literary work we are required to submit to the force and agency of that work and so made aware that intention, for all its apparent elusiveness and mystery, was there in the text all along: as its desire, its organizing principle, both its form and that form's undoing.

Note

1 A recent and interesting example of the attribution of cognitivity to text can be found in Michael Wood's article on *The Wings of the Dove* in the *London Review of Books* (2003), where he asks what the text 'knows', and implies that a work of literature can know things that its creator may not know or have known; he uses James's novel to explore the relationship between literature and knowledge, and to speculate on what sorts of 'knowledge' the text itself might demonstrate.

Bibliography

Aagaard-Mogensen, L. and L. De Vos (eds). 1986. *Text, Literature, & Aesthetics* (Amsterdam: Rodopi)

Aiken, H. D. 1955. 'The Aesthetic Relevance of the Artist's Intentions', *Journal of Philosophy*, 52.24: 742–53

Altieri, Charles. 1975. 'The Poem as Act', *Iowa Review*, 6.3: 103–24

—1981. *Act & Quality: A Theory of Literary Meaning & Humanistic Understanding* (Brighton: Harvester Press)

—1994. *Subjective Agency: A Theory of First-Person Expressivity & Its Social Implications* (Oxford: Blackwell)

Anscombe, G. E. 1963. *Intention* (second edn) (Cambridge, MA: Harvard University Press)

Aristotle. 1962. *On Interpretation*, trans. by J. T. Oesterle, *Medieval Texts in Translation No 11* (Milwaukee: Marquette University Press)

—1996. *Poetics*, trans. by M. Heath (Harmondsworth: Penguin Books)

Austin, J. L. 1962. *How to Do Things with Words* (Oxford: Oxford University Press)

—1963. 'Performative-Constative', in Searle (ed.). *The Philosophy of Language* (1971) (London: Oxford University Press), pp. 13–22

Avni, O. 1990. *The Resistance of Reference* (Baltimore: Johns Hopkins University Press)

Ayer, A. J. 1971. *Language, Truth & Logic* (Harmondsworth: Penguin Books)

Barthes, Roland. 1977. *Image Music Text*, trans. by S. Heath (London: Fontana Press)

Beardsley, Monroe C. 1958. *Aesthetics: Problems in the Philosophy of Criticism* (New York: Harcourt, Brace & World, Inc.)

—1965. 'On the Creation of Art', in Beardsley and Schueller (eds). *Aesthetic Inquiry* (1967) (Belmont: Dickenson Publishing Company), pp. 171–87

—1968. 'Textual Meaning & Authorial Meaning', *Genre*, 1.3: 169–81

—1970. *The Possibility of Criticism* (Detroit: Wayne State University Press)

—1973. 'The Concept of Literature', in Brady, Palmer and Price (eds). *Literary Theory & Structure* (New Haven: Yale University Press), pp. 23–39

—1978a. 'Aesthetic Intentions & Fictive Illocutions', in Hernadi (ed.). *What is Literature?* (Bloomington: Indiana University Press), pp. 161–77

—1978b. 'Intending', in Goldman and Kim (eds). *Values & Morals* (Dordrecht: D. Reidel Publishing Co.), pp. 163–84

—1981. 'Fiction as Representation', *Synthese*, 46.3: 291–313

—1982. *The Aesthetic Point of View: Selected Essays*, ed. by M. J. Wreen and D. M. Callen (Ithaca: Cornell University Press)

Beardsley, Monroe C. and H. M. Schueller (eds). 1967. *Aesthetic Inquiry* (Belmont: Dickenson Publishing Company)

Bearn, G. C. F. 1995. 'Derrida Dry: Iterating Iterability Analytically', *Diacritics*, 25.3: 3–25
Belsey, Catherine. 1982. 'Problems of Literary Theory: The Problem of Meaning', *New Literary History*, 14.1: 175–82
Benjamin, Walter. 1999. *Illuminations*, trans. by H. Zorn (London: Pimlico)
Benoist, J.-M. 1978. *The Structural Revolution* (London: Weidenfeld & Nicholson)
Berthoff, A. E. 1982a. 'I. A. Richards & the Audit of Meaning', *New Literary History*, 14.1: 63–79
—1982b. 'Response to Catherine Belsey', *New Literary History*, 14.1: 185–86
Biriotti, M. and N. Miller (eds). 1993. *What is an Author?* (Manchester: Manchester University Press)
Black, M. (ed.). 1965. *Philosophy in America* (London: Allen & Unwin)
—1973. 'Meaning & Intention: An Examination of Grice's Views', *New Literary History*, 4.2: 257–79
—1979. 'More About Metaphor', in Ortony (ed.). *Metaphor & Thought* (Ithaca: Cornell University Press), pp. 19–43
Blanchot, Maurice. 1982. *The Space of Literature*, trans. by A. Smock (Lincoln: University of Nebraska Press)
Boly, J. R. 1985. 'Nihilism Aside: Derrida's Debate Over Intentional Models', *Philosophy & Literature*, 9.2: 152–65
Booth, Wayne. 1988. *The Company We Keep* (Berkeley: University of California Press)
Brady, F., J. Palmer and M. Price (eds). 1973. *Literary Theory & Structure* (New Haven: Yale University Press)
Brentano, Franz. 1969. *The Origin of Our Knowledge of Right & Wrong*, trans. by R. M. Chisholm and E. H. Schneewind (London: Routledge)
Brooks, C. 1951. 'The Formalist Critics', *The Kenyon Review*, 13.1: 72–81
Brooks, Peter. 1984. *Reading for the Plot: Design & Intention in Narrative* (Cambridge, MA: Harvard University Press)
Brown, M. E. 1968. 'Interpretation of Poetry', *Genre*, 1.3: 230–41.
Brown Jr, R. L. and M. Steinmann Jr. 1978. 'Native Readers of Fiction: A Speech-Act & Genre-Rule Approach to Defining Literature', in Hernadi (ed.). *What is Literature?* (Bloomington: Indiana University Press), pp. 141–60
Burke, K. 1975. 'Words as Deeds', *Centrum*, 3.2: 147–68
Burke, Séan. 1995. *Authorship: From Plato to the Postmodern, A Reader* (Edinburgh: Edinburgh University Press)
—1998. *The Death & Return of the Author* (second edn) (Edinburgh: Edinburgh University Press)
Burkhardt, A. (ed.) 1990. *Speech Acts, Meanings & Intentions* (Berlin: De Gruyter)
Cain, W. E. 1977. 'Authority, "Cognitive Atheism", & the Aims of Interpretation: The Literary Theory of E. D. Hirsch', *College English*, 39.3: 333–45
—1980. 'Authors & Authority in Interpretation', *Georgia Review*, 34.3: 617–34
Carrithers, G. H. 1968. 'How Literary Things Go: Contra Hirsch', *Genre*, 1.3: 195–208
Carroll, D. 1982. *The Subject in Question* (Chicago: Chicago University Press)
Carroll, Noel. 1992. 'Art, Intention, & Conversation', in Iseminger (ed.). *Intention & Interpretation* (Philadelphia: Temple University Press), pp. 97–131

Cascardi, Anthony. (ed.). 1987. *Literature & the Question of Philosophy* (Baltimore: Johns Hopkins University Press)
Chatman, Seymour. (ed.) 1971. *Literary Style: A Symposium* (Oxford: Oxford University Press)
—1973. *Approaches to Poetics* (New York: Columbia University Press)
Chisholm, Roderick. 1970. 'The Structure of Intention', *Journal of Philosophy*, 67.19: 633–47
Cioffi, Frank. 1976. 'Intention & Interpretation in Criticism', in Newton-de Molina (ed.). *On Literary Intention*, pp. 55–73. Originally published in *Proceedings of the Aristotelian Society*, 64 (1963–64): 85–106
Cixous, Helene. 1991. *Coming to Writing*, ed by D. Jenson, trans. by S. Cornell, D. Jenson, A. Liddle and S. Sellers (Cambridge, MA: Harvard University Press)
Close, A. J. 1976. 'Don Quixote & the Intentional Fallacy', in Newton-de Molina (ed.). *On Literary Intention*, pp. 174–93. Originally published in *British Journal of Aesthetics*, 12.1 (1972): 19–39
Crewe, J. 1985. 'Toward Uncritical Practice', in Mitchell (ed.). *Against Theory*, pp. 53–64. Originally published in *Critical Inquiry*, 9.4 (1983): 748–59
Critchley, Simon. 1992. *Ethics of Deconstruction* (Oxford: Blackwell)
Culler, Jonathan. 1975. *Structuralist Poetics* (London: Routledge)
—1983. *On Deconstruction* (London: Routledge)
—1997. *Literary Theory: A Very Short Introduction* (Oxford: Oxford University Press)
Cunningham, Valentine. 2002. *Reading After Theory* (Oxford: Blackwell)
Currie, Gregory. 1985. 'What is Fiction?', *Journal of Aesthetics & Art Criticism*, 43.4: 385–92
—1986. 'Works of Fiction & Illocutionary Acts', *Philosophy & Literature*, 10.2: 304–08
Dasenbrock, Reed Way (ed.). 1989. *Redrawing the Lines: Analytic Philosophy, Deconstruction, & Literary Theory* (Minneapolis: University of Minnesota Press)
—1993. *Literary Theory After Davidson* (Pennsylvania: Penn State University Press)
Davies, D. 1999. 'Artistic Intentions & the Ontology of Art', *British Journal of Aesthetics*, 39.2: 148–62
Davis, W. 1984. 'A Causal Theory of Intending', *American Philosophical Quarterly*, 21.1: 43–54
Davis, W. A. 1998. *Implicature* (Cambridge: Cambridge University Press)
de Man, Paul. 1978. 'Foreword', in C. Jacobs (ed.). *The Dissimulating Harmony* (Baltimore: Johns Hopkins University Press), pp. vii–xiii
—1979. *Allegories of Reading* (New Haven: Yale University Press)
—1983. *Blindness & Insight: Essays in the Rhetoric of Contemporary Criticism* (second edn) (London: Routledge)
—1986. *The Resistance to Theory* (Minneapolis: University of Minnesota Press)
Derrida, Jacques. 1973. *Speech & Phenomena*, trans. by D. B. Allison (Evanston: Northwestern University Press)
—1976. *Of Grammatology*, trans. by G. Spivak (Baltimore: Johns Hopkins University Press)
—1978. *Writing & Difference*, trans. by A. Bass (London: Routledge)
—1981a. *Dissemination*, trans. by B. Johnson (London: The Athlone Press)
—1981b. *Positions*, trans. by A. Bass (London: The Athlone Press)
—1982. *Margins of Philosophy*, trans. by A. Bass (London: Prentice Hall)

—1988. *Limited Inc*, trans. by S. Weber (Evanston: Northwestern University Press)
Dickie, G. 1968. 'Meaning & Intention', *Genre*, 1.3: 182–89
Dowling, W. C. 1985. 'Intentionless Meaning', in Mitchell (ed.). *Against Theory*, pp. 89–94. Originally published in *Critical Inquiry*, 9.4 (1983): 784–89
Dutton, Denis. 1987. 'Why Intentionalism Won't Go Away', in Cascardi (ed.). *Literature & the Question of Philosophy* (Baltimore: Johns Hopkins University Press), pp. 192–209
Eaglestone, Robert. 1997. *Ethical Criticism: Reading After Levinas* (Edinburgh: Edinburgh University Press)
Eagleton, Terry. 1996. *Literary Theory: An Introduction* (second edn) (Oxford: Blackwell)
—2003. *After Theory* (London: Allen Lane)
Earnshaw, Steven. 1996. *The Direction of Literary Theory* (Basingstoke: Macmillan Press)
Eaton, M. 1969. 'Art, Artifacts & Intentions', *American Philosophical Quarterly*, 6.2: 165–69
Efron, A. 1968. 'Logic, Hermeneutic, & Literary Context', *Genre*, 1.3: 214–29
Ellis, A. J. 1974. 'Intention & Interpretation in Literature', *British Journal of Aesthetics*, 14.4: 315–25
Farrell, F. B. 1988. 'Iterability & Meaning: The Searle–Derrida Debate', *Metaphilosophy*, 19.1: 53–64
Fish, Stanley. 1978. 'Normal Circumstances, Literal Language, Direct Speech Acts, the Ordinary, the Everyday, the Obvious, What Goes without Saying, & Other Special Cases', *Critical Inquiry*, 4.4: 625–44
—1980. *Is There a Text in this Class?* (Cambridge, MA: Harvard University Press)
—1989. *Doing What Comes Naturally* (Oxford: Clarendon Press)
Fish, Stanley, Barbara Herrnstein Smith, E. D. Hirsch and M. Steinmann Jr. 1975. 'Discussion', *Centrum*, 3.2: 125–46
Fisher, J. (ed.). 1983. *Essays on Aesthetics* (Philadelphia: Temple University Press)
Fleming, B. N. 1964. 'On Intention', *Philosophical Review*, 73.3: 301–20
Fodor, Jerry. 2002. 'Mouse Thoughts', *London Review of Books*, 24.5: 12–14
Foucault, Michel. 1970. *The Order of Things* (London: Routledge)
—1977a. 'What Is An Author?', in D. F. Bouchard (ed. and trans.). *Language, Counter-Memory, Practice* (Ithaca: Cornell University Press), pp. 113–38
—1977b. 'Nietzsche, Genealogy, History', in *Language, Counter-Memory, Practice* (Ithaca: Cornell University Press), pp. 139–64
Fowler, A. 1976. 'Intention Floreat', in Newton-de Molina (ed.). *On Literary Intention* (Edinburgh: Edinburgh University Press), pp. 242–55
Frye, Northrop. 1957. *Anatomy of Criticism* (Harmondsworth: Penguin Books)
Gadamer, Hans-Georg. 1989. *Truth & Method* (second edn) trans. by J. Weinsheimer and D. G. Marshall (London: Sheed & Ward)
Gang, T. M. 1957. 'Intention', *Essays in Criticism*, 7.2: 175–86
Gasché, Rodolphe. 1986. *The Tain of the Mirror* (Cambridge, MA: Harvard University Press)
—1987. 'Infrastructures & Systematicity', in Sallis (ed.). *Deconstruction & Philosophy* (Chicago: University of Chicago Press), pp. 3–20
Gendin, S. 1964. 'The Artist's Intentions', *Journal of Aesthetics & Art Criticism*, 23.2: 193–96

Gibbs, R. W. 1999. *Intentions in the Experience of Meaning* (Cambridge: Cambridge University Press)

Glendinning, Simon. 1998. *On Being With Others* (London: Routledge)

Goldman, A. I. 1970. *A Theory of Human Action* (Englewood Cliffs: Prentice-Hall)

Goldman, A. I. and J. Kim (eds). 1978. *Values & Morals* (Dordrecht: D. Reidel)

Grice, H. P. 1969. 'Utterer's Meaning & Intentions', *Philosophical Review*, 78.2: 147–77

—1971a. 'Intention & Uncertainty', *Proceedings of the British Academy*, 57: 263–79

—1971b. 'Utterer's Meaning, Sentence-Meaning, & Word-Meaning', in Searle (ed.). *The Philosophy of Language*, pp. 54–70. Originally published in *Foundations of Language*, 4 (1968): 225–42

—1989a. 'Meaning', in *Studies in the Way of Words* (Cambridge, MA: Harvard University Press), pp. 213–23. Originally published in *Philosophical Review*, 66.3 (1957): 377–88

—1989b. 'Logic & Conversation', in *Studies in the Way of Words* (Cambridge, MA: Harvard University Press), pp. 22–40. Originally published in P. Cole and J. L. Morgan (eds). *Syntax & Semantics: Vol. 3: Speech Acts* (New York: Academic Press, 1975), pp. 41–58

—1989c. 'Meaning Revisited', in *Studies in the Way of Words* (Cambridge, MA: Harvard University Press), pp. 283–303

Hamlyn, D. W. 1971. 'Unconscious Intentions', *Philosophy*, 46.1: 12–22

Hancher, Michael. 1972. 'Three Kinds of Intention', *Modern Language Notes*, 87.3: 827–51

—1975. 'Understanding Poetic Speech Acts', *College English*, 36.6: 632–39

—1977. 'Beyond a Speech Act Theory of Literary Discourse', *Modern Language Notes*, 92.4: 1081–98

—1979. 'The Classification of Cooperative Illocutionary Acts', *Language in Society*, 8.1: 1–14

—1981. 'Humpty Dumpty & Verbal Meaning', *Journal of Aesthetics and Art Criticism*, 40.1: 49–58

Hanfling, Oswald. 1996. 'Fact, Fiction & Feeling', *British Journal of Aesthetics*, 36.4: 356–66

Harold, J. 2000. 'Empathy with Fictions', *British Journal of Aesthetics*, 40.3: 340–55

Hartman, Geoffrey. 1981. *Saving the Text: Literature/Derrida/Philosophy* (Baltimore: Johns Hopkins University Press)

Heidegger, Martin. 1971. *On the Way to Language*, trans. by P. D. Hertz (New York: Harper & Row)

Hermerén, G. 1983. 'The Autonomy of Art', in Fisher (ed.). *Essays on Aesthetics* (Philadelphia: Temple University Press), pp. 35–49

Hernadi, P. (ed.) 1978. *What is Literature?* (Bloomington: Indiana University Press)

Hirsch, E. D. 1967. *Validity in Interpretation* (New Haven: Yale University Press)

—1975. 'What's the Use of Speech-Act Theory?', *Centrum*, 3.2: 121–24

—1976. *The Aims of Interpretation* (Chicago: Chicago University Press)

—1977. *The Philosophy of Composition* (Chicago: Chicago University Press)

—1984. 'Meaning & Significance Reinterpreted', *Critical Inquiry*, 11.2: 202–25

—1985. 'Against Theory?', in Mitchell (ed.). *Against Theory* (Chicago: Chicago University Press), pp. 48–52. Originally published in *Critical Inquiry*, 9.4 (1983): 743–47

—1994. 'Transhistorical Intentions & the Persistence of Allegory', *New Literary History*, 25.3: 549–67

Howells, C. 1999. *Derrida: Deconstruction from Phenomenology to Ethics* (Cambridge: Polity Press)

Hungerland, Isobel C. 1955. 'The Concept of Intention in Art Criticism', *Journal of Philosophy*, 52.24: 733–42

Huntley, J. 1968. 'A Practical Look at E. D. Hirsch's *Validity in Interpretation*', *Genre*, 1.3: 242–55

Husserl, Edmund. 2001a. *Logical Investigations*, Vol. 1, trans. by J. N. Findlay, ed. by D. Moran (London: Routledge)

—2001b. *Logical Investigations*, Vol. 2, trans. by J. N. Findlay, ed. by D. Moran (London: Routledge)

Hutcheon, Linda 1988. *A Poetics of Postmodernism: History, Theory, Fiction* (London: Routledge)

Ingarden, Roman. 1973a. *The Cognition of the Literary Work of Art*, trans. by Ruth Ann Crowley and Kenneth R. Olson (Evanston: Northwestern University Press)

—1973b. *The Literary Work of Art*, trans. by George G. Grabowicz (Evanston: Northwestern University Press)

Irwin, William. 1999. *Intentionalist Interpretation: A Philosophical Explanation & Defence* (Connecticut: Greenwood Press)

Iseminger, Gary. 1992a. *Intention & Interpretation* (Philadelphia: Temple University Press)

—1992b. 'An Intentional Demonstration?', in *Intention & Interpretation*, pp. 76–96

—1996. 'Actual Intentionalism vs. Hypothetical Intentionalism', *Journal of Aesthetics & Art Criticism*, 54.4: 319–26

—1998. 'Interpretive Relevance, Contradiction, & Compatibility with the Text: A Rejoinder to Knight', *Journal of Aesthetics & Art Criticism*, 56.1: 58–61

Iser, Wolfgang. 1974. *The Implied Reader* (Baltimore: Johns Hopkins University Press)

—1978. *The Act of Reading* (Baltimore: Johns Hopkins Univeristy Press)

—2006. *How to Do Theory* (Oxford: Blackwell)

Jakobson, Roman. 1987. *Language in Literature* (Cambridge, MA.: Harvard University Press)

Jancovich, Mark. 1993. *The Cultural Politics of the New Criticism* (Cambridge: Cambridge University Press)

Johnson, C. 1993. *System & Writing in the Philosophy of Jacques Derrida* (Cambridge: Cambridge University Press)

—1997. *Derrida: The Scene of Writing* (London: Orion)

Jones, H. M. 1964. 'The Relevance of the Artist's Intentions', *British Journal of Aesthetics*, 4.1: 138–45

Joyce, R. 2000. 'Rational Fear of Monsters', *British Journal of Aesthetics*, 40.2: 209–24

Juhl, P. D. 1980. *Interpretation: An Essay in the Philosophy of Literary Criticism* (Guildford: Princeton University Press)

Kamuf, Peggy. 1988. *Signature Pieces* (Ithaca: Cornell University Press)

Kemp, J. 1964. 'The Work of Art & the Artist's Intentions', *British Journal of Aesthetics*, 4.1: 146–54

Kent, T. 1989. 'Beyond System: The Rhetoric of Paralogy', *College English*, 51.5: 492–507

Knapp, Steven. 1993. *Literary Interest: The Limits of Anti-Formalism* (Cambridge, MA: Harvard University Press)
Knapp, Steven and Walter Benn Michaels. 1985a. 'Against Theory', in Mitchell (ed.). *Against Theory* (Chicago: Chicago University Press), pp. 11–30. Originally published in *Critical Inquiry*, 8.4 (1982): 723–42
—1985b. 'A Reply to Our Critics', in Mitchell (ed.). *Against Theory* (Chicago: Chicago University Press), pp. 95–105. Originally published in *Critical Inquiry*, 9.4 (1983): 790–800
—1985c. 'A Reply to Richard Rorty: What is Pragmatism?', in Mitchell (ed.). *Against Theory* (Chicago: Chicago University Press), pp. 139–45. Originally published in *Critical Inquiry*, 11.3 (1985): 466–73
—1992. 'The Impossibility of Intentionless Meaning', in Iseminger (ed.). *Intention and Interpretation* (Philadelphia: Temple University Press), pp. 51–64. This article combines material originally published as part of 'Against Theory', in Mitchell (ed.). *Against Theory* (1985), pp. 11–30, and 'Against Theory 2', *Critical Inquiry*, 14.1 (1987): 49–68.
—1994. 'Reply to John Searle', *New Literary History*, 25.3: 669–75
Knight, D. 1998. 'Not an Actual Demonstration: A Reply to Iseminger', *Journal of Aesthetics and Art Criticism*, 56.1: 53–58
Konigsberg, I. (ed.) 1981. *American Criticism in the Poststructuralist Age* (Ann Arbor: University of Michigan Press)
Krausz, M. 1992. 'Intention & Interpretation: Hirsch & Margolis', in Iseminger (ed.). *Intention & Interpretation* (Philadelphia: Temple University Press), pp. 152–66
Kripke, Saul. 1980. *Naming & Necessity* (Oxford: Blackwell)
Kuhns, R. 1960. 'Criticism & the Problem of Intention', *Journal of Philosophy*, 57.1: 5–23
Lamarque, Peter. 1990. 'The Death of the Author: An Analytical Autopsy', *British Journal of Aesthetics*, 30.4: 319–31
—1996. *Fictional Points of View* (Ithaca: Cornell University Press)
Leddy, T. 1999. 'Iseminger's Literary Intentionalism & an Alternative', *British Journal of Aesthetics*, 39.3: 219–29
Lentricchia, Frank. 1983. *After the New Criticism* (London: Methuen)
LePore, Ernest. 1983. 'The Concept of Meaning & Its Role in Understanding Language', *Dialectica*, 37.2: 133–39
Levin, S. R. 1976. 'Concerning What Kind of Speech Act a Poem Is', in van Dijk (ed.). *Pragmatics of Language & Literature* (Amsterdam: North Holland Publishing Co.), pp. 141–60
Livingston, P. 1991. *Literature & Rationality: Ideas of Agency in Theory & Fiction* (Cambridge: Cambridge University Press)
—1996. 'Arguing Over Intentions', *Revue International de Philosophie*, 4: 615–33
—2005. *Art & Intention: A Philosophical Study* (Oxford: Clarendon)
Llewelyn, J. 1986. *Derrida on the Threshold of Sense* (Basingstoke: Macmillan Press)
Lodge, David. (ed.). 1988. *Modern Criticism & Theory* (London: Longman)
Lyas, Colin. 1973. 'Personal Qualities & the Intentional Fallacy', in *Philosophy & the Arts: Royal Institute of Philosophy Lectures* Vol. 6, 1971–72 (Basingstoke: Macmillan Press), pp. 194–210

—1983. 'Anything Goes: The Intentional Fallacy Revisited', *British Journal of Aesthetics*, 23.4: 291–305
—1986. 'Speech Acts & the Personal Heresy', in Aagaard and De Vos (eds). *Text, Literature, & Aesthetics* (Amsterdam: Rodopi), pp. 37–65
—1992. 'Wittgensteinian Intentions', in Iseminger (ed.). *Intention & Interpretation* (Philadelphia: Temple University Press), pp. 132–51
McGonigal, A. 2002. 'Metaphor, Indeterminacy, & Intention', *British Journal of Aesthetics*, 42.2: 179–90
McKenna, W. R. and J. Claude Evans (eds). 1995. *Derrida & Phenomenology* (Dordrecht: Kluwer)
Macksey, Richard and Eugenio Donato (eds). 1970. *The Structuralist Controversy: The Languages of Criticism & the Sciences of Man* (Baltimore: Johns Hopkins University Press)
Maier, R. 1970. '"The Intentional Fallacy" & the Logic of Literary Criticism', *College English*, 32.2: 135–45
Mailoux, S. 1985. 'Truth or Consequences: On Being Against Theory', in Mitchell (ed.). *Against Theory* (Chicago: Chicago University Press), pp. 65–71. Originally published in *Critical Inquiry*, 9.4 (1983): 760-66
Margolis, Jonathan. (ed.). 1962. *Philosophy Looks at the Arts* (New York: Scribner)
— 1979. 'Literature & Speech Acts', *Philosophy & Literature*, 3.1: 39–52
—1992. 'Robust Relativism', in Iseminger (ed.). *Intention & Interpretation* (Philadelphia: Temple University Press), pp. 41–50. Originally published in Jonathan Margolis, *Art & Philosophy: Conceptual Issues in Aesthetics* (Atlantic Highlands: Humanities Press, 1980), pp. 156–64
—1995. *Interpretation Radical but Not Unruly* (Berkeley: University of California Press)
McQuillan, M. 2001. *Paul de Man* (London: Routledge)
Matejka, L. and K. Pomorska (eds). 1978. *Readings in Russian Poetics: Formalist & Structuralist Views* (Ann Arbor: University of Michigan Press)
Meiland, Jack W. 1978. 'Interpretation as a Cognitive Discipline', *Philosophy & Literature*, 2.1: 23–45
—1981. 'The Meanings of a Text', *British Journal of Aesthetics*, 21: 195–203
Miller, J. Hillis 1987. *The Ethics of Reading* (New York: Columbia University Press)
—2001. *Speech Acts in Literature* (Stanford: Stanford University Press)
Mitchell, W. J. T. (ed.) 1985. *Against Theory: Literary Studies & the New Pragmatism* (Chicago: Chicago University Press)
Moran, Dermot. 2000. *Introduction to Phenomenology* (London: Routledge)
Nathan, D. O. 1982. 'Irony & the Artist's Intentions', *British Journal of Aesthetics*, 22.3: 245–56
—1992. 'Irony, Metaphor, & the Problem of Intention', in Iseminger (ed.). *Intention & Interpretation* (Philadelphia: Temple University Press), pp. 183–202
Nehamas, Alexander. 1981. 'The Postulated Author', *Critical Inquiry*, 8.1: 133–49
—1986. 'What an Author Is', *Journal of Philosophy*, 83.11: 685–91
—1987. 'Writer, Text, Work, Author', in Cascardi (ed.). *Literature & the Question of Philosophy* (Baltimore: Johns Hopkins University Press), pp. 265–91
Newton, K. M. (ed.). 1997. *Twentieth Century Literary Theory: A Reader* (second edn) (Basingstoke: Macmillan Press)

Newton-de Molina, D. (ed.). 1976. *On Literary Intention* (Edinburgh: Edinburgh University Press)
Norris, Christopher. 1985. *The Contest of Faculties* (London: Methuen)
—1988. *Paul de Man: Deconstruction & the Critique of Aesthetic Ideology* (London: Routledge)
—1990. *What's Wrong with Postmodernism* (Baltimore: Johns Hopkins University Press)
—1992. *Deconstruction & the Interests of Theory* (Leicester: Leicester University Press)
—1997. *Resources of Realism* (Basingstoke: Macmillan Press)
—2002. *Deconstruction: Theory & Practice* (third edn) (London: Routledge)
Ohmann, Richard. 1966. 'Literature as Sentences', *College English*, 27.4: 261–67
—1971a. 'Speech, Action, & Style', in Chatman (ed.). *Literary Style: A Symposium* (Oxford: Oxford University Press), pp. 241–54
—1971b. 'Speech Acts & the Definition of Literature', *Philosophy & Rhetoric*, 4.1: 1–19
—1972. 'Speech, Literature & the Space Between', *New Literary History*, 4.1: 47–63
—1973. 'Literature as Act', in Chatman (ed.). *Approaches to Poetics* (New York: Columbia University Press), pp. 81–107
—1978. 'The Social Definition of Literature', in Hernadi (ed.). *What is Literature?* (Bloomington: Indiana University Press), pp. 89–101
Olsen, C. 1969. 'Knowledge of One's Own Intentional Actions', *Philosophical Quarterly*, 19.77: 324–36
Olsen, Stein Haugom. 1973. 'Authorial Intention', *British Journal of Aesthetics*, 13.3: 219–31
—1977. 'Interpretation & Intention', *British Journal of Aesthetics*, 17.3: 210–18
—1978. *The Structure of Literary Understanding* (Cambridge: Cambridge University Press)
—1982a. 'The "Meaning" of a Literary Work', *New Literary History*, 14.1: 13–32
—1982b. 'Reply', *New Literary History*, 14.1: 187–89
Osborne, H. 1972. *Aesthetics* (Oxford: Oxford University Press)
Oxford English Dictionary (second edn) Vol. VII (Oxford: Clarendon Press, 1989)
Pappas, N. 1989. 'Authorship & Authority', *Journal of Aesthetics & Art Criticism*, 47.4: 325–32
Parker, H. 1985. 'Lost Authority: Non-Sense, Skewed Meanings & Intentionless Meanings', in Mitchell (ed.). *Against Theory* (Chicago: Chicago University Press), pp. 72–79. Originally published in *Critical Inquiry*, 9.4 (1983): 767–74
Payzant, G. 1964. 'Intention & the Achievement of the Artist', *Dialogue*, 3.2: 154–56
Peckham, Morse. 1968. 'Semantic Autonomy & Immanent Meaning', *Genre*, 1.3: 190–94
—1976. 'The Intentional? Fallacy?', in Newton-de Molina (ed.). *On Literary Intention* (Edinburgh: Edinburgh University Press), pp. 139–57
Petrey, Sandy. 1990. *Speech Acts & Literary Theory* (London: Routledge)
Plato. 1973. *Phaedrus & The Seventh & Eighth Letters*, trans. by W. Hamilton (Harmondsworth: Penguin Books)
—1974. *The Republic* (second edn) trans. by D. Lee (Harmondsworth: Penguin Books)
Poulet, Georges. 1964. *The Interior Distance*, trans. by E. Coleman (Ann Arbor: University of Michigan Press)

—1969. 'Phenomenology of Reading', *New Literary History*, 1.1: 53–68
—1970. 'Criticism & the Experience of Interiority', in Macksey and Donato (eds). *The Structuralist Controversy* (Baltimore: Johns Hopkins University Press), pp. 56–72
Pradhan, S. 1986. 'Minimalist Semantics: Davidson & Derrida on Meaning, Use, & Convention', *Diacritics*, 16.1: 66–77
Pratt, Mary-Louise. 1977. *Toward a Speech Act Theory of Literary Discourse* (Bloomington: Indiana University Press)
—1981. 'The Ideology of Speech-Act Theory', *Centrum* New Series 1.1: 5–18
Price, C. 2003. 'Artificial Functions & the Meaning of Literary Works', *British Journal of Aesthetics*, 43.1: 1–17
Radford, Colin. 2002. 'How Can We Be Moved by the Fate of Anna Karenina?', in A. Neill and A. Ridley (eds). *Arguing About Art* (London: Routledge), pp. 239–49
Ray, William. 1984. *Literary Meaning: From Phenomenology to Deconstruction* (Oxford: Blackwell)
Recanati, Francois. 1987. *Meaning & Force* (Cambridge: Cambridge University Press)
Redpath, T. 1976. 'The Meaning of a Poem', in Newton-de Molina (ed.). *On Literary Intention* (Edinburgh: Edinburgh University Press), pp. 14–25. Originally published in C. A. Mace (ed.). *British Philosophy in the Mid-Century* (London: Allen & Unwin, 1957), pp. 361–75
Richmond, S. 1996. 'Derrida & Analytical Philosophy: Speech Acts & Their Force', *European Journal of Philosophy*, 4.1: 38–62
Ricoeur, Paul. 1973. 'The Model of the Text: Meaningful Action Considered as a Text', *New Literary History*, 5.1: 91–117
Rockas, L. 1968. 'Validity in Evaluation', *Genre*, 1.3: 209–13
Roma III, E. 1976. 'The Scope of the Intentional Fallacy', in Newton-de Molina (ed.). *On Literary Intention* (1976) (Edinburgh: Edinburgh University Press), pp. 74–86. Originally published in *The Monist*, 50.2 (1966): 250–66
Rorty, Richard. 1982. *Consequences of Pragmatism* (Minneapolis: University of Minnesota Press)
—1985. 'Philosophy without Principles', in Mitchell (ed.). *Against Theory* (Chicago: Chicago University Press), pp. 132–38. Originally published in *Critical Inquiry*, 11.3 (1985): 459–65
Rosebury, B. 1997. 'Irrecoverable Intentions & Literary Interpretation', *British Journal of Aesthetics*, 37.1: 15–30
Roskill, Mark. 1977. 'On the "Intention" & "Meaning" of Works of Art', *British Journal of Aesthetics*, 17.2: 99–110
Ryle, Gilbert. 1949. *The Concept of Mind* (Harmondsworth: Penguin Books)
Sallis, J. (ed.). 1987. *Deconstruction & Philosophy* (Chicago: University of Chicago Press)
de Saussure, Ferdinand. 1959. *Course in General Linguistics*, trans. by W. Baskin (New York: McGraw-Hill Book Company)
Savile, Anthony. 1972. 'The Place of Intention in the Concept of Art', in Osborne (ed.). *Aesthetics* (Oxford: Oxford University Press), pp. 158–76. First published in *Proceedings of the Aristotelian Society*, 69 (1968–69): 101–24
—1998. 'Imagination & the Content of Fiction', *British Journal of Aesthetics*, 38.2: 136–49
Searle, John. 1965. 'What is a Speech Act?', in Black (ed.). *Philosophy in America* (London: Allen & Unwin), pp. 221–39

—1969. *Speech Acts* (Cambridge: Cambridge University Press)
Searle, John (ed.). 1971. *The Philosophy of Language* (Oxford: Oxford University Press)
—1977. 'Reiterating the Differences: A Reply to Derrida', *Glyph*, 1: 198–208
—1979. *Expression & Meaning* (Cambridge: Cambridge University Press)
Searle, John, F. Kiefer and M. Bierwisch. 1980. *Speech Act Theory & Pragmatics* (Dordrecht: D. Reidel)
—1983. *Intentionality: An Essay in the Philosophy of Mind* (Cambridge: Cambridge University Press)
—1994a. 'Literary Theory & Its Discontents', *New Literary History*, 25.3: 637–67
—1994b. 'Structure & Intention in Language: A Reply to Knapp & Michaels', *New Literary History*, 25.3: 677–81
—1999. *Mind, Language & Society* (London: Weidenfeld & Nicholson)
Shapiro, G. 1974. 'Intention & Interpretation in Art: A Semiotic Analysis', *Journal of Aesthetics & Art Criticism*, 33.1: 33–42
Shklovsky, V. 1988. 'Art as Technique', in Lodge (ed.). *Modern Criticism & Theory* (London: Longman), pp. 15–30
Shusterman, Richard. 1984. *The Object of Literary Criticism* (Amsterdam: Rodopi)
—1992. 'Interpretation, Intention, & Truth', in Iseminger (ed.). *Intention & Interpretation* (Philadelphia: Temple University Press), pp. 65–75. Originally published in *Journal of Aesthetics & Art Criticism*, 46.3 (1988): 399–411
Siebers, Tobin. 1988. *The Ethics of Criticism* (Ithaca: Cornell University Press)
Sircello, G. 1986. 'Monroe Beardsley & American Aesthetics', in Aagaard-Mogensen and De Vos (eds). *Text, Literature, & Aesthetics* (Amsterdam: Rodopi), pp. 111–22
Skinner, Quentin. 1976. 'Motives, Intentions & the Interpretation of Texts', in Newton-de Molina (ed.). *On Literary Intention* (Edinburgh: Edinburgh University Press), pp. 210–21. Originally published in *New Literary History*, 3.2 (1972): 393–408.
Smith, B. and D. W. Smith (eds). 1995. *Cambridge Companion to Husserl* (Cambridge: Cambridge University Press)
Smith, Barbara Herrnstein 1975. 'Actions, Fictions, & the Ethics of Interpretation', *Centrum*, 3.2: 117–20
—1978. *On the Margins of Discourse: The Relation of Literature to Language* (Chicago: University of Chicago Press)
—1988. *Contingencies of Value: Alternative Perspectives for Critical Theory* (Cambridge, MA: Harvard University Press)
Smith, P. 1988. *Discerning the Subject* (Minneapolis: University of Minnesota Press)
Sokolowski, R. 2000. *Introduction to Phenomenology* (Cambridge: Cambridge University Press)
Sparshott, F. E. 1967. *The Concept of Criticism* (Oxford: Oxford University Press)
—1976. 'Criticism & Performance' in Newton-de Molina (ed.). *On Literary Intention* (Edinburgh: Edinburgh University Press), pp. 104–15. Originally published in *The Concept of Criticism* (1967), pp. 162–77.
—1986. 'The Case of the Unreliable Author', *Philosophy & Literature*, 10.2: 145–67
Spurlin, W. and M. Fischer (eds). 1995. *The New Criticism & Contemporary Literary Theory: Connections & Continuities* (New York: Garland Publishing Inc.)
Staten, Henry. 1993. 'The Secret Name of Cats: Deconstruction, Intentional Meaning, & the New Theory of Reference', in Dasenbrock (ed.). *Literary Theory After Davidson* (Pennsylvania: Penn State University Press), pp. 27–48

Stecker, Robert. 1987. 'Apparent, Implied, & Postulated Authors', *Philosophy & Literature*, 11.2: 258–71

Steig, M. 1977. 'The Intentional Phallus', *Journal of Aesthetics & Art Criticism*, 36.1: 51–61

Steinmann Jr, M. 1975. 'Perlocutionary Acts & the Interpretation of Literature', *Centrum*, 3.2: 112–16

Stout, J. 1982a. 'What is the Meaning of a Text?', *New Literary History*, 14.1: 1–12

—1982b. 'The Problem of Meaning: A Reply to Belsey', *New Literary History*, 14.1: 183–84

Strawson, P. F. 1964. 'Intention & Convention in Speech Acts', *Philosophical Review*, 73.4: 439–60

Swinden, Patrick. 1999. *Literature & the Philosophy of Intention* (Basingstoke: Macmillan Press)

Tillyard, E. M. W. and C. S. Lewis. 1939. *The Personal Heresy: A Controversy* (Oxford: Oxford University Press)

Tolhurst, W. E. 1979. 'On What a Text Is & How It Means', *British Journal of Aesthetics*, 19.1: 3–14

Trivedi, S. 2001. 'An Epistemic Dilemma for Actual Intentionalism', *British Journal of Aesthetics*, 41.2: 192–206

van Dijk, T. A. 1976. *Pragmatics of Language & Literature* (Amsterdam: North Holland Publishing Co.)

van Peer, W. 1989. *The Taming of the Text* (London: Routledge)

Walton, K. 1978. 'Fearing Fictions', *Journal of Philosophy*, 75: 5–27

—1983. 'Fiction, Fiction-Making, & Styles of Fictionality', *Philosophy & Literature*, 7.1: 78–88

—1993. *Mimesis of Make-Believe* (Cambridge, MA: Harvard University Press)

Wellek, Rene and A. Warren. 1963. *Theory of Literature* (third edn) (Harmondsworth: Penguin Books)

Wheeler III, Samuel. C. 2000. *Deconstruction as Analytic Philosophy* (Stanford: Stanford University Press)

White, A. 1995. 'Of Grammatolatry: Deconstruction as Rigorous Phenomenology?', in McKenna and Evans (eds). *Derrida & Phenomenology* (Dordrecht: Kluwer), pp. 103–19

Williams, Bernard. 1985. *Ethics & the Limits of Philosophy* (London: Fontana Press)

—1995. 'Ethics', in Grayling (ed.). *Philosophy, A Guide Through the Subject*, pp. 545–82

Wimsatt, W. K. 1954. *The Verbal Icon* (Lexington: University of Kentucky Press)

Wimsatt, W. K. and Monroe C. Beardsley. 1962. 'The Intentional Fallacy', in Margolis (ed.). *Philosophy Looks at the Arts*, pp. 91–105. Originally published in *Sewanee Review*, 54.3 (1946): 468–88

—1968. 'Genesis: An Argument Resumed', in *Day of the Leopards* (1976), pp. 11–39

—1976. *Day of the Leopards* (New Haven: Yale University Press)

Wittgenstein, Ludwig. 1953. *Philosophical Investigations*, trans. by G. E. M. Anscombe (Oxford: Blackwell)

Wood, Michael. 2003. 'What Henry Knew', *London Review of Books*, 25.24: 21–25

INDEX

www.ingramcontent.com/pod-product-compliance
Ingram Content Group UK Ltd.
Pitfield, Milton Keynes, MK11 3LW, UK
UKHW020144290926
15471UKWH00004B/682

9 781441 198648